T0135773

# ENZYMATIC STABILIZATION AND FORMATION OF FOOD NANO- AND MICROSTRUCTURES

**Dissertation zur Erlangung des Doktorgrades**

**der Naturwissenschaften (Dr. rer. nat.)**

**Fakultät Naturwissenschaften**

**Universität Hohenheim**

Institut für Lebensmittelwissenschaft und Biotechnologie

vorgelegt von

*Benjamin Zeeb*

geboren am 27.02.1982 in Nürtingen

Bibliografische Information der Deutschen Nationalbibliothek

Die Deutsche Nationalbibliothek verzeichnet diese Publikation in der Deutschen Nationalbibliografie; detaillierte bibliografische Daten sind im Internet über http://dnb.d-nb.de abrufbar.



ISBN 978-3-8325-3922-1

Logos Verlag Berlin GmbH
Comeniushof, Gubener Str. 47,
10243 Berlin
Tel.: +49 (0)30 42 85 10 90
Fax: +49 (0)30 42 85 10 92
INTERNET: http://www.logos-verlag.de

Dekan: Prof. Dr. Heinz Breer
Institut für Physiologie
Universität Hohenheim

1. berichtende Person: Prof. Dr. Jochen Weiss
Institut für Lebensmittelwissenschaft und Biotechnologie
Universität Hohenheim

2. berichtende Person: Prof. Dr. Lutz Fischer
Institut für Lebensmittelwissenschaft und Biotechnologie
Universität Hohenheim

Eingereicht am: 20.09.2013

Mündliche Prüfung am: 07.03.2014

Die vorliegende Arbeit wurde am *17.03.2014* von der Fakultät Naturwissenschaften der Universität Hohenheim als „Dissertation zur Erlangung des Doktorgrades der Naturwissenschaften" angenommen.

# Coauthors

The scientific work presented was partially conducted in corporation with other scientists working at the Department of Food Physics and Meat Science, University of Hohenheim. Prof. Dr. Jochen Weiss was supervising the whole PhD thesis.

- CHAPTER 2: Benjamin Zeeb planned the study. PhD Monika Gibis helped in performing the HPLC experiments used to determine the ferulic acid content of sugar beet pectin.

- CHAPTER 3 and 4: Benjamin Zeeb planned the study and was also responsible for the experimental work. Statistical issues were solved by PhD Monika Gibis.

- CHAPTER 5: Benjamin Zeeb planned the study. The experimental work presented in this chapter was carried out by Johanna Beicht.

- CHAPTER 6: The study was planned by Benjamin Zeeb and PhD Hanna Salminen. She assisted in the preparation of the biopolymer nanoparticles presented in this chapter.

- CHAPTER 7: Benjamin Zeeb planned the study together with Thomas Eisele (Department of Food Biotechnology). He also conducted the transglutaminase activity test presented in this chapter. Johanna Beicht partly carried out the crosslinking experiments.

# List of publications

Parts of this thesis have been previously published with the approval of the supervisor.

**Peer-reviewed publications**

Beicht, J., Zeeb, B., Gibis, M., Fischer, L. and Weiss, J., *Influence of layer thickness and composition of cross-linked multilayered oil-in-water emulsions on the release behavior of lutein*. Food & Function, 2013. DOI:10.1039/C3FO60220F.

Zeeb, B., Zhang, H., Gibis, M., Fischer, L. and Weiss, J., *Influence of buffer on the preparation of multilayered emulsions stabilized by proteins and polysaccharides*. Food Research International, 2013. 53(1): p. 325-333.

Zeeb, B., Fischer, L. and Weiss, J., *Hofmeister salts affect the buildup of thin multilayer films surrounding oil droplets*. Journal of Dispersion Science and Technology, 2013. DOI:10.1080/01932691.2013.813395.

Zeeb, B., Gibis, M., Fischer, L. and Weiss, J., *Influence of interfacial properties on Ostwald ripening in crosslinked multilayered oil-in-water emulsions*. Journal of Colloid and Interface Science, 2012. 387(1): p. 65-73.

Zeeb, B., Gibis, M., Fischer, L. and Weiss, J., *Crosslinking of interfacial layers in multilayered oil-in-water Emulsions using laccase: Characterization and pH-Stability*. Food Hydrocolloids, 2012. 27(1): p. 126-136.

Kessler, A., Zeeb, B., Kranz, B. Menéndez-Aguirre, O., Fischer, L., Hinrichs, J. and Weiss, J., *Isothermal titration calorimetry as a tool to determine thermodynamics of demicellization processes*. Review of Scientific Instruments, 2012. 83(10): Art-no 105104.

Zeeb, B., Fischer, L. and Weiss, J., *Cross-Linking of interfacial layers affects the salt and temperature stability of multilayered emulsions consisting of fish gelatin and sugar beet pectin*. Journal of Agricultural and Food Chemistry, 2011. 59(19): p. 10546–10555

**Oral presentations**

Zeeb, B., Fischer, L., and Weiss, J., *Biomimetic approach to stabilize and functionalize food nano- and microstructures*. Delivery of Functionality in Complex Food Systems, Physically-Inspired Approaches from the Nanoscale to the Microscale, 5th International Symposium, 2013. Haifa, Israel.

Zeeb, B., Fischer, L., and Weiss, J., *Ostwald Reifung in enzymbehandelten mehrschichtigen Emulsionen*. Processnet: Jahrestreffen der Fachausschüsse Lebensmittelverfahrenstechnik, Rheologie und Trocknungstechnik, 2012. Hohenheim, Germany.

Zeeb, B., Fischer, L., and Weiss, J., *Enzymatische Stabilisierung von Nano- und Mikrostrukturen von Lebensmitteln*. Processnet: Jahrestreffen des Fachausschusses Lebensmittelverfahrenstechnik, 2011. Vlaardingen, The Netherlands.

Zeeb, B., Fischer, L., and Weiss, J., *Vernetzung von Grenzflächenmembranen in mehrschichtigen Emulsionen mit Laccase*. Wissenschaftliches Kolloquium Food Sciene and Biotechnology, 2011. Hohenheim, Germany.

**Poster presentations**

Maier, C., Zeeb, B., and Weiss, J., *Screening of oppositely charged food- and cosmetic grade O/W emulsions for controlled heteroaggregation*. Delivery of Functionality in Complex Food Systems, Physically-Inspired Approaches from the Nanoscale to the Microscale, 5th International Symposium, 2013. Haifa, Israel.

Zeeb, B., Salminen, H., and Weiss, J., *Biomimetic approach to stabilize biopolymer particles formed by complex coacervation*. Dairy Conference, 2013. Hohenheim, Germany.

Zeeb, B., Beicht, J., Gibis, M., Fischer, L., and Weiss, J., *Laccase-induced crosslinking of multilayered oil-in-water emulsions impacts the release behavior of encapsulated lutein*. IFT Annual Meeting and Food Expo, 2013. Chicago, USA.

Guldiken, B., Boyacioglu, D., Sramék, M., Zeeb, B., Gibis, M., Kohlus, R., and Weiss, J., *Characterization of spray drying ability of WPI/Beet pectin coacervates stabilized oil-in-water emulsions*. IFT Annual Meeting and Food Expo, 2013. Chicago, USA.

Beicht, J., Zeeb, B., Gibis, M., Fischer, L., and Weiss, J., *Influence of layer thickness and composition of multilayered oil-in-water emulsions on the release behavior of lutein*. ELLS Scientific Student Conference – Future Cities – Future Life, 2012. Alnarp, Schweden.

Zeeb, B., Gibis, M., Fischer, L., und Weiss, J., *Beeinflussung des Lutein-Releases durch Modifizierung der Membranstruktur mehrschichtiger Öl-in-Wasser Emulsionen*. GDL-Kongress Lebensmitteltechnologie, 2012. Dresden, Germany.

Zeeb, B., Fischer, L., and Weiss, J., *Improving electrostatic deposition of biopolymer layers on emulsion interfaces by addition of salts*. 16$^{th}$ IUFoST World Congress of Food Science and Technology, 2012. Foz do Iguaçu, Brazil.

Zeeb, B., Gibis, M., Fischer, L., and Weiss, J., *Ostwald ripening in multilayered oil-in-water emulsions*. IFT Annual Meeting and Food Expo, 2012. Las Vegas, USA.

Zeeb, B., Fischer, L., and Weiss, J., *Enzymatic crosslinking of interfacial membranes improves functionality of multilayered emulsions*. Delivery of Functionality in Complex Food Systems; Physically-Inspired Approaches from the Nanoscale to the Microscale, 4th International Symposium, 2011. Guelph, Canada.

Zeeb, B., Gibis, M., Fischer, L., and Weiss, J., *Laccase-induced crosslinking of interfacial layers in multilayered oil-in-water emulsions*. Delivery of Functionality in Complex Food Systems; Physically-Inspired Approaches from the Nanoscale to the Microscale, 4th International Symposium, 2011. Guelph, Canada.

# Abstract

Food technologists and food manufacturers are increasingly asked to develop tailor-made delivery systems that are able to perform a number of different roles in the food system, such as for example provide for an enhanced physical and chemical stability of an encapsulated compound, and to control the release location and kinetics of a compound. The most frequently used food delivery systems include solid particle suspensions, simple and multiple emulsions, gels, solid matrices, and association colloids. Unfortunately, with the exception of association colloids, most of these delivery systems are thermodynamically unstable, i.e. they have a tendency to destabilize and phase separate over time. This is because the integrity of the assembled structures depends on a large number of physical interaction forces between colloids and molecules, such as Van der Waals, electrostatic, hydrophobic, depletion, and hydration interactions. These physical interaction forces are highly dependent upon environmental conditions such as pH, salt concentrations, and temperatures. While this provides food scientist with a larger toolbox to assemble structures, it is also a substantial drawback since any change in environmental condition may alter the carefully crafted structures. In this thesis, a new approach to the stabilization of food structures - namely the use of crosslinking enzymes - was investigated in depth. A number of different enzymes known to crosslink various types of biopolymers was used to this purpose. An example of one such enzymes is the enzyme transglutaminase, an acyltransferase that has shown to covalently crosslink a free amine group on a protein with a $\gamma$-carboxamid group on another protein. Transglutaminase has already been used commercially in a variety of food applications including the production of meat, dairy, fish, and cereal products, albeit in a relatively uncontrolled manner as far as structure modification on a colloidal or nanoscalar level is concerned. Recently, enzymes belonging to the class of oxidases have been suggested to also be able to act as crosslinking agents in food applications. One such promising enzyme belonging to this class is for example laccase. To test the hypothesis that such crosslinking enzymes are able to fixate a structure created initially via physical interaction forces and thereby provide the needed long term stability, studies were conducted using model food dispersions as targets of enzymatic crosslinking with systems having well defined structures on a molecular and colloidal level. Specifically, the effect of addition of laccase as crosslinking enzyme on biopolymer layers or biopolymer aggregates adsorbed at the interface of dilute and concentrated oil-in-water emulsions was investigated. Results of the studies presented in this thesis demonstrate that laccase added to multilayered or aggregate-stabilized oil-in-water emulsions composed of a model triacylglyceride oil and fish gelatin and sugar beet pectin as interfacial biopolymers induced crosslinks in sequentially deposited interfaces thereby improving the stability of the generated structures against superimposed stresses (pH, salt addition, freeze-thaw cycling, thermal processing). In contrast, laccase was unable to crosslink biopolymer aggregates adsorbed at the surface of emulsion droplets due to spatial restrictions that physically limit access of the enzymes to the substrate. Interfacial composition, layer thickness, and layer density affected the effectiveness of the crosslinking approach to reduce Ostwald ripening - a key destabilizing mechanism in emulsions composed of short chain lipids. Moreover, enzymatic treatment led to alterations in release of lutein - a hydrophobic bioactive compound that we use as a model bioactive compound – from biopolymer-coated emulsions. Finally, transglutaminase promoted the formation of protein-stabilized emulsion gels when the oil volume fraction exceeded > 0.6. Taken together, the studies presented in this thesis demonstrate that crosslinking enzymes have a substantial potential to stabilize food dispersions against a great variety of common destabilization mechanisms as long as the enzymes are not physically restricted in their access to the target substrate. Use of enzymatic crosslinking thus offers an environmentally-friendly and economically feasible approach to rationally design novel delivery and encapsulation systems for the food industry.

# Aim of the study

In this thesis project, the use of enzymes to modify and thereby stabilize food structures assembled by means of physical interaction forces was investigated. Our base hypothesis was that the formation of covalent bonds catalyzed by the biochemical action of enzymes may make the dispersion less susceptible to changes in environmental conditions thereby providing them with better long term stability. As such, the project's aim was to replicate the action of nature, which in many cases uses the same sequence of first physically assembling and later enzymatically stabilizing biologically important structures. The goal of this work was to establish a fundamental understanding as to how food nano- and microstructures that are currently of great interest to the food industry as encapsulation systems are affected by the action of crosslinking-capable enzymes. Of particular interest was the question as to whether the spatial arrangement of a target structure has an influence on its ability to be crosslinked and how the stability and functional properties of the final system are altered. Ultimately, we wanted to know whether this approach may be of use to food manufacturers to generate new products. To this purpose, a number of questions were asked such as:

I. Is an enzymatic-induced crosslinking of biopolymers present in or adsorbed to a carrier system generally possible? (CHAPTER 2)

II. Does a crosslinking of biopolymers in a particular carrier system enhance its functionality in terms of stability and release properties? (CHAPTER 3, 4, 5)

III. Does the spatial distribution of biopolymers in the respective structure affect the ability of enzyme to access the target? (CHAPTER 6)

IV. Can this approach to be used to rationally derive complex structures from base template structures? (CHAPTER 6, 7)

Each of these questions led to one or more studies that were in turn submitted for review and publication to established journals in the field of food and colloid science.

# Symbols and abbreviations

## Symbols

| | | |
|---|---|---|
| $a^*$ | color coordinate: red green direction | |
| $b^*$ | color coordinate: yellow blue direction | |
| $c(\infty)$ | bulk solubility | mol/m$^3$ |
| $c(r)$ | solubility of a compound in the immediate vicinity of a particle of radius $r$ | mol/m$^3$ |
| $c_a$ | mass of emulsifier adsorbed to the surface of the droplets per unit volume emulsion | kg/m$^3$ |
| $c_i$ | initial mass of emulsifier per volume | kg/m$^3$ |
| $c_w$ | mass of emulsifier in the aqueous phase | kg/m$^3$ |
| $D$ | droplet diameter | m |
| $D_0$ | diffusion coefficient for the dispersed phase molecules in the continuous phase | m$^2$/s |
| $d_{10}$ | number-length mean diameter | m |
| $d_{32}$ | volume-surface mean droplet diameter | m |
| $d_{43}$ | volume fraction-length mean diameter | m |
| $G'$ | storage modulus | Pa |
| $G''$ | loss modulus | Pa |
| $H$ | mean surface distance | m |
| $I$ | ionic strength | M |
| $k_B$ | Boltzmann constant ($1.38 * 10^{-23}$) | J/K |
| $L^*$ | color coordinate: lightness | |
| $M_w$ | molecular weight | kg/mol |
| $n_i$ | number of droplets | |
| p$I$ | isoelectric point | |
| p$K_a$ | acid dissociation constant | |
| $R$ | molar gas constant | J/mol*K |
| $R$ | radius of a droplet | m |
| $r_c$ | critical radius of a droplet | m |
| $T$ | absolute temperature | K |
| $T$ | time | min |

| | | |
|---|---|---|
| $\tan \delta$ | loss tangent | |
| $V_m$ | molar volume of the dispersed phase | $m^3/mol$ |
| $\gamma^i$ | interfacial tension | N/m |
| $\Gamma_S$ | surface load | $kg/m^2$ |
| $H$ | dynamic viscosity | $N*s/m^2$ |
| $\eta^*$ | complex viscosity | Pa*s |
| $P$ | density | $kg/m^3$ |
| $\tau_{OR}$ | characteristic time of Ostwald ripening | s |
| $\Omega$ | Ostwald ripening rate | $m^3/s$ |
| $\Phi$ | oil volume fraction | |

**Abbreviations**

| | |
|---|---|
| (1) | primary emulsion |
| (2) | secondary emulsion |
| (2+) | enzymatically-treated secondary emulsion |
| AMD | age-related macular degeneration |
| AU | activity units |
| AUA | galacturonic acid monohydrate |
| BHT | butylatedhydroxytoluene |
| $CaCl_2$ | calcium chloride |
| Da | Dalton |
| DLS | dynamic light scattering |
| DTAB | dodecyltrimethylammonium bromide |
| EC number | Enzyme Commission number |
| FA | ferulic acid |
| FG | fish gelatin |
| GDL | glucono-δ-lactone |
| HCl | hydrochloric acid |
| HLB | hydrophilic-lipophilic-balance |
| HPLC | high performance liquid chromatography |
| ITC | isothermal titration calorimetry |
| LbL | layer-by-layer electrostatic depositioning |
| LSW theory | Lifshitz-Slyozov-Wagner theory |

| | |
|---|---|
| M | molar per liter |
| Min | minute(s) |
| MLVs | multilamellar vesicles |
| Na-Cas | sodium caseinate |
| NaCl | sodium chloride |
| NaOH | sodium hydroxide |
| NIR | near infrared |
| O/W | oil-in-water emulsion |
| OR | Ostwald ripening |
| PC | phosphatidylcholin (lecithin) |
| PDI | polydispersity index |
| PEG | polyethylene glycol |
| PPO | polyphenoloxidase |
| PSD | particle size distribution |
| Psi | pound-force per square inch (≈ 6895 Pa) |
| Rpm | revolutions per minute |
| SDS | sodium dodecylsulfate |
| SLN | solid lipid nanoparticle |
| SLS | static light scattering |
| TAG | triacylglyceride |
| TCA | trichloroacetic acid |
| TGase | microbial transglutaminase |
| TNBS | 2,4,6-Trinitrobenzenesulfonic |
| ULVs | unilamellar vesicles |
| V | volume |
| W | weight |
| W/O | water-in-oil emulsion |
| WPI | whey protein isolate |
| z-Ave | mean particle diameter (= z-average) |
| Z-Gln-Gly | Z-glutamic acid-glycine |

# Table of contents

# CHAPTER 1

## Stabilization of food dispersions by enzymes

*Benjamin Zeeb*[1], *Monika Gibis*[1], *Lutz Fischer*[2], *Jochen Weiss*[1]

[1] Department of Food Physics and Meat Science, University of Hohenheim, Garbenstrasse 21/25, 70599 Stuttgart, Germany

[2] Department of Food Biotechnology, University of Hohenheim, Garbenstrasse 25,70599 Stuttgart, Germany

Adapted from "*Stabilization of food dispersions using enzymes*", Zeeb, B., Fischer, L., Weiss, J., Food Function, 2014, 5(2), p. 198-213 with permission from The Royal Society of Chemistry.

## ABSTRACT

Food dispersions have become essential vehicles to carry and deliver functional ingredients such as bioactive compounds, flavors, antimicrobials, antioxidants, colors, and vitamins. Most of these systems are thermodynamically unstable tending to break down over time. Much research has therefore been carried out to develop methodologies to improve their long-term stability. In this review, we will introduce readers to a new approach that has been developed over the past years to stabilize food dispersions, i.e. by the use of various enzymes. First, basic design principles of modern food dispersions including conventional emulsions, multiple emulsions, multilayered emulsions, solid lipid particle suspensions, and liposomes are being discussed. Enzymes able to generate intra- and intermolecular crosslinks between proteins and/or polysaccharides will be reviewed and specific reactions catalyzed by e.g. transglutaminase, laccase, tyrosinase, sulfhydryl oxidase, glucose oxidase, lipoxygenase, polyphenol oxidase, peroxidase, and lysyl oxidase will be highlighted. Finally, potential applications of this enzymatic approach in the food industry will be critically discussed.

**Keywords:** Functional Foods, Encapsulation, Delivery Systems, Emulsions, Micelles, Coacervates, Self-Assembly, Enzyme, Laccase, Transglutaminase

## INTRODUCTION

The food industry is currently experiencing a major paradigm shift. Instead of simply demanding sufficient quantities of safe food that satisfy the caloric intake needs of the population, consumer increasingly ask that their food also promote their health and wellbeing (*Mota et al.*, 2000; *Sanguansri et al.*, 2006). Modern health and wellness promoting foods are commonly known as functional or novel foods and contain additional ingredients (so called nutraceuticals) that have shown to affect human physiology in a positive way (*Aguilera*, 2006; *Chen et al.*, 2006; *Manski et al.*, 2007; *Weiss et al.*, 2008). The list of bioactive or nutraceutical components that may be added to foods is continuously growing. Carotenoids, omega-3 fatty acids, vitamins, bioactive peptides, antioxidants, minerals, modified fats are just a few examples of components that are being assessed for or are already used in the fortification of foods. The inclusion of these compounds in foods, however, has proven to be extremely challenging due to the fact that these compounds (a) are often chemically labile and prone to undergo degradation reactions such as oxidation, (b) have a tendency to interact with the multitude of ingredients that are typically present in complex foods and, (c) lose their efficacy when incorporated in foods (*Chen et al.*, 2006; *Weiss et al.*, 2008). The challenge for food technologists is therefore to incorporate these bioactive ingredients into various food matrices without negatively affecting their organoleptic properties, to maintain their chemical integrity in the food matrix prior to consumption, and to ensure that compounds are truly physiologically effective when ingested. To date, the most suitable and economical feasible way to incorporate these components in food products is to encapsulate them in nano- or microencapsulation systems (*Desai et al.*, 2005; *Weiss et al.*, 2006; *McClements et al.*, 2009).

Bioactive compounds differ in their molecular properties such as molecular weight, structures, functional groups, polarities, conformations and charge. As a result, their physicochemical properties such as solubility, physical state, rheology, surface activity vary greatly (*McClements*, 2006; *Dickinson*, 2008). Therefore, food technologists and food manufacturers are required to develop very specific, tailor-made delivery systems that comply with a number of key performance requirements. For example, delivery systems are required to (1) contain an appreciable amount of a functional component, (2) protect the bioactive component against chemical degradation such as oxidation or hydrolysis, (3) allow for a controlled-release of the functional ingredients at an appropriate time and at an appropriate location, (4) be compatible with the specific surrounding matrix, and (5) be physically stable, that is resist a large number of environmental stresses that may be superimposed during production, storage, and transport. To this purpose, a large number of delivery and encapsulation systems have been created by food scientists to incorporate bioactive and functional ingredients (*McClements et al.*, 2007; *McClements et al.*, 2009) (**Figure 1**).

Frequently used delivery systems include suspensions, biopolymer particles, simple and multiple emulsions, gels, liposomal vesicles, solid particles, and association colloids (*Schmitt et al.*, 1998; *McClements*, 2004; *Taylor et al.*, 2005; *McClements et al.*, 2007; *Branco et al.*, 2009; *Maherani et al.*, 2011; *Matalanis et al.*, 2011; *McClements et al.*, 2011). Most of these are carrier systems can be easily dispersed in an aqueous phase allowing for a precise dosing of the functional ingredient. Aqueous dispersability is also important because (a) a majority of food systems contains as prime "ingredient" water, and (b) highly concentrated formulations can be shipped at low costs to food processors around the world and diluted to the required concentration on site saving substantial costs.

Unfortunately, with the exception of the association colloids, most of these delivery systems are thermodynamically unstable, i.e. they have a tendency to destabilize and phase separate over time (*Dickinson et al.*, 1999). For the production and stabilization of dispersed systems such as emulsions or suspensions, a variety of different emulsifiers such as proteins (caseins, whey proteins) or polysaccharides (pectin, gum arabic, modified starch) are used. The composition, charge, density and thickness of the interfacial membrane of an oil-in-water emulsion determines not only its stability, but in many cases affects also the product's texture, appearance, shelf life, taste, and aroma (*Guzey et al.*, 2006).

Emulsifiers may vary greatly in their molecular, physical and chemical characteristics and are therefore more or less well suited to stabilize food dispersions. For every individual application an appropriate emulsifier type must therefore be selected. In general, the selection of food dispersion constituents is a very difficult task and requires an intrinsic knowledge not only of the to be created food dispersion, but also of the matrix to which the food dispersion is to be added, the process involved in the fabrication of the intermediate and final product, the shelf life that is to be achieved and the potential stresses the intermediate or final product is subject to. Fluctuations in conditions along this fabrication and use chain may occur, affecting at any given time the quality of the end product.

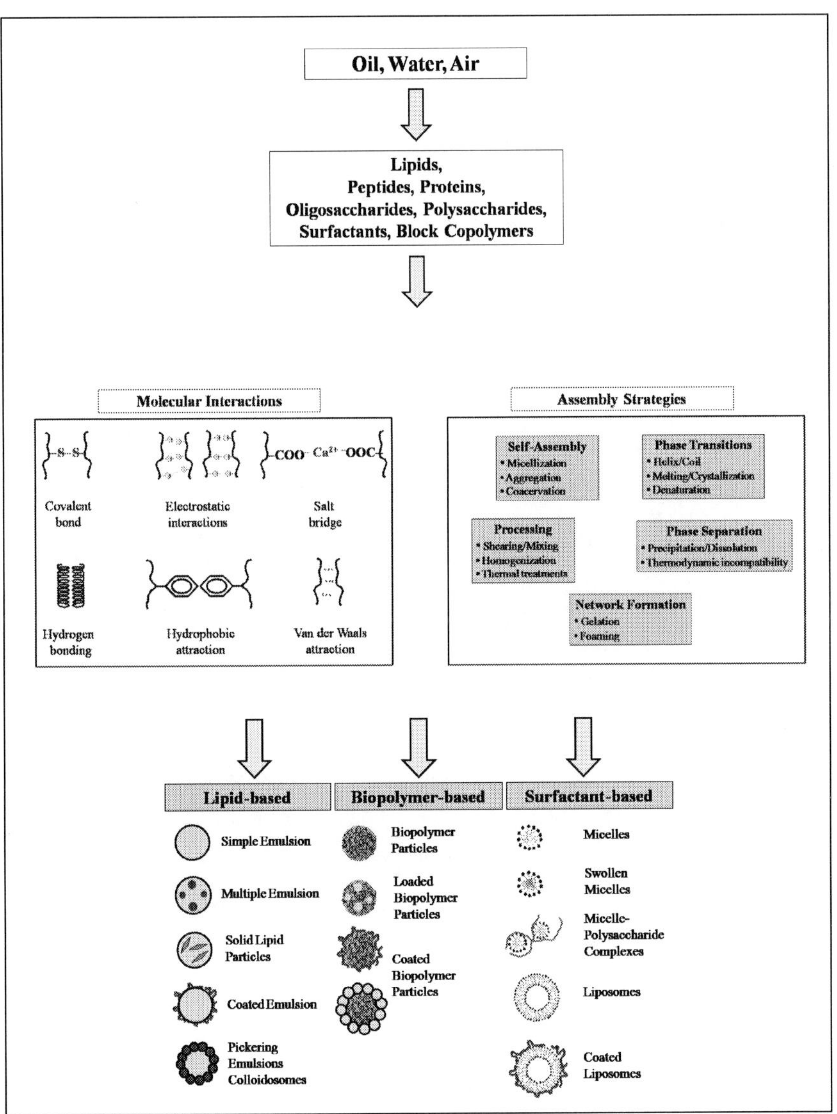


**Figure 1 Structural design principles to create self-assembled delivery systems which are in need of enzymatic stabilization (*Branco et al.*, 2009; *McClements et al.*, 2009).**

Aside from the selection of appropriate ingredients, food scientists have to select suitable assembly techniques to build the food dispersion. For example, dispersions can be generated by mechanical means such as the use of milling, grinding, mixing, and homogenization processes or they can be structured by use of physical phenomena such as phase separation, phase transition, aggregation, and coacervation – to name just a few (*McClements et al.*, 2007; *McClements et al.*, 2009). Once generated, the structures are subject to molecular or colloidal interactions such as Van der Waals, electrostatic, hydrophobic, depletion, and hydration interactions. These interactions are highly dependent on environmental conditions such as pH, salt concentrations, and temperatures and may irrevocably alter the carefully crafted structure of the food dispersion (*McClements*, 2004; *McClements et al.*, 2009). An in-depth understanding of these interactions is therefore needed to ensure that once the food dispersion has been manufactured, its structure can be maintained for the duration of the use.

An approach to overcome this issue is to employ a variety of physical, chemical, and/or enzymatic means that induce chemical crosslinks rather than just relying on physical interaction bonds to stabilize the generated structures (*Dickinson*, 1997; *Gerrard*, 2002; *Kato*, 2002). To some degree, such an approach mimics biological processes, in which molecules are first assembled by means of physical interactions and then crosslinked to fixate them. In a broader sense, this is a part of the newly developing science of biomimicry, where processes from nature are translated to generate innovations in engineering, design or computing (*Estroff et al.*, 2001; *Estroff*, 2008).

This review highlights in particular the use of enzymes to modify and stabilize assembled food structures. Previous studies in food biotechnology have indicated that the modification and stabilization of assemblies of natural polymers such as proteins and polysaccharides by enzyme technology is an economically feasible option, and that a wide variety of food enzymes approved for the use in foods are available that could be used to this purpose.

## TARGET STRUCTURES

The following section gives a brief overview of the various food dispersions that could profit from an enzymatic stabilization. All of these structures can be potentially used as delivery and carrier systems by the food industry to encapsulate nutraceuticals und functional components. Most of these carrier systems are emulsions-type systems that can be easily dispersed in an aqueous phase. In general, many food products partly or wholly consist of emulsions which explains their dominance amongst food dispersions (*McClements*, 2004; *Appelqvist et al.*, 2007) (**Figure 1**).

**Conventional emulsions**

An emulsion can be defined as a dispersion of two completely or partially immiscible liquids, usually of oil and water. Emulsions can be categorized into two main classes according to the relative spatial distribution of the oil and water phase: (i) dispersions of water droplets in oil are referred to as water-in-oil (w/o) emulsions, and (ii) dispersions of oil droplets in water are referred to as oil-in-water (o/w) emulsions (*Davis et al.*, 1987; *McClements*, 2004). Emulsions with more complex spatial distributions can also be produced for example emulsions in which the dispersed droplets themselves consist again of emulsions. These types of emulsions are called multiple or double emulsions. Examples of such emulsions are the so called WOW emulsions, where water droplets are dispersed in oil droplets that are again dispersed in a water phase (w/o/w), and the so called OWO emulsions, where oil droplets are dispersed in water droplets that are again dispersed in an oil phase (o/w/o) (*Yazan et al.*, 1993; *Garti*, 1998; *Garti et al.*, 1998; *Muschiolik*, 2007). The high interfacial area between the different phases is a result of the dispersion of one phase in the other, and is energetically highly unfavourable. Since the two liquids involved in the formation have very different polarities (one being hydrophobic, the other being hydrophilic), there is no enthalpy gained from mixing them, and enthalpy and entropy losses make the creation of emulsions thus highly unfavourable. Increases in free energy are directly proportional to the increase in interfacial area with the surface tension being the proportionality factor (*McClements*, 2004; *Rodríguez-Abreu et al.*, 2008). To reduce the surface tension and the magnitude of the free energy change, surface active substances (surfactants, emulsifiers) are used. These substances are major contributors to enhancing the stability of emulsions. Processes such as gravitational separation (e.g. creaming or sedimentation), droplet aggregation (e.g. flocculation or coalescence), and Ostwald ripening are responsible for the breakdown of emulsions over time (*Dalgleish*, 1997; *McClements et al.*, 2009). Selection of suitable formulations and process conditions are currently the main approach to prevent their occurrence.

For the manufacturing of conventional oil-in-water (o/w) emulsions oil, water, an emulsifier, and mechanical energy input is needed; a process commonly known as homogenization (*Dickinson*, 1997). During homogenization, oil droplets are disrupted and new interfacial areas are generated. Emulsifier molecules adsorb to these freshly formed interfaces lowering their interfacial tension. As a consequence, oil droplets which are surrounded by a thin layer of emulsifier molecules are formed and dispersed in an aqueous continuous phase as illustrated in **Figure 1**. The most commonly used emulsifiers in the food industry are amphiphilic proteins, polysaccharides, phospholipids and small molecule surfactants (*Guzey et al.*, 2006).

**Multiple emulsions**

Multiple emulsions such as water-in-oil-in-water (w/o/w) emulsions consist of small water droplets that are present within larger oil droplets that are dispersed in an aqueous phase (*Garti*, 1997; *Garti et al.*, 1998; *Benichou et al.*, 2004) (**Figure 1**). These kinds of emulsions are of great interest to the food industry to encapsulate or protect sensitive and bioactive ingredients such as flavors, vitamins, minerals, enzymes or to produce foods with lower fat content (*McClements et al.*, 1998; *Yaqoob Khan et al.*, 2006; *McClements et al.*, 2007; *Muschiolik*, 2007). The production of multiple emulsions is more complex than that of simple emulsions. Typically, a two-step method is used whereby a primary o/w emulsions is first prepared under high shear conditions, while the secondary emulsification step is carried out under mild shear conditions. Double emulsions consist of two interfacial layers (**Figure 1**); consequently, two different types of emulsifiers are needed to kinetically stabilize these emulsions. The stabilization of these systems has proven to be quite challenging (*Garti*, 1998). Multiple emulsions tend to break down during storage or when external stresses such as mechanical forces, thermal strains, cooling or freezing are superimposed.

In WOW emulsions, functional food components may be encapsulated within the inner water phase, the oil phase, or be dispersed in the outer water phase, thereby making it possible to develop a single delivery system that contains multiple functional components. This technology could be used physically separate two water-soluble components that might adversely react with each other if they were present in the same aqueous phase – an example of such component combinations system would be Vitamin C and iron. Alternatively, one could use such systems to release component sequentially at specific sites such as first the mouth, then the stomach, or small intestine (*Weiss et al.*, 2006).

**Oil bodies**

Recently, oil bodies extracted from soybeans have gained attention due to potential applications in foods, cosmetics, and pharmaceuticals (*Peng et al.*, 2003; *Chen et al.*, 2004). These lipid storage organelles are naturally found in the seeds of many plants seeds (*Murphy et al.*, 2000; *Murphy et al.*, 2001; *Shimada et al.*, 2010). Oil bodies consist of a lipid core that is surrounded by a phospholipid-olesin layer (*Tzen et al.*, 1992; *Tzen et al.*, 1993). The olesin layer forms a natural barrier in order to protect the oil bodies against environmental stresses such as moisture and temperature changes, and the presence of reactive oxygen species. Oil bodies could be used as pre-emulsified oils in food products such as dressings, sauces, dips, beverages, and desserts to improve stability during food processing, storage, transport, and utilization (*Iwanaga et al.*, 2007; *Iwanaga et al.*, 2008).

**Micelles and microemulsions**

Amphiphilic molecules having hydrophilic head and hydrophobic tail groups spontaneously self-assemble under appropriate environmental conditions (e.g. temperature, ionic strength, and pH) to form so called association colloids including micelles, bilayers, and reversed micelles (*Weiss et al.*, 2009) (**Figure 1**). The molecular properties of the surfactant itself (e.g. geometry of head and tail group, polarity, and charge), the properties of the solvent (e.g. pH, ionic strength, and dielectric constant), the presence of any co-surfactants, as well as the overall (co-)surfactant concentration, and the temperature determine the type and structure of association colloid formed (*McClements et al.*, 2009). The association of amphiphilic molecules is usually driven by the hydrophobic effect which causes the system to minimize all unfavourable contact areas between the non-polar tails of surfactant molecules and the aqueous solvent (*McClements et al.*, 2009). Surfactant molecules are present as monomers in solution and tend to spontaneously self-assemble into thermodynamically stable aggregates if the concentration exceeds a critical value, commonly known as the critical micelle concentration (CMC) (*Weiss et al.*, 2006). Furthermore, micelles may organize themselves into crystalline structures such as hexagonal, lamellar, and reversed hexagonal phases, at higher surfactant concentrations (*McClements et al.*, 2009). The size and shape of micelles is determined by the molecular geometry of the surfactants, solvent composition, and temperature (*McClements et al.*, 2009).

Surfactant micelles have a number of remarkable properties, such as thermodynamic stability, aqueous solubility (i.e. their surfaces are fully polar), high mobility, and optical transparency due to lack of interaction with light waves having wavelengths in the range of visible light. However, large quantities of surfactant molecules are required to assemble them leading to cost, off-flavor, and potentially toxicity problems. Dilutability is also an issue, since their structure is concentration dependent. At high concentrations, gel-like phases may be formed which are difficult to dilute. On the other hand, if they are diluted to concentrations below the CMC, dissociation takes place (*Weiss et al.*, 2009).

Association colloids can be used to incorporate and deliver both polar and nonpolar functional components (*Garti et al.*, 2005; *Flanagan et al.*, 2006). Lipophilic ingredients may be solubilized into the hydrophobic core of the micelle or be a part of the membrane structure (the palisade layer) resulting in the formation of core-shell surfactant-lipid particles, better known as "swollen micelles" or "microemulsions" (*McClements et al.*, 2009; *Weiss et al.*, 2009). The formation of a microemulsion, however, is simple and cost effective requiring only mixing of components. Microemulsions vary in particle sizes between 5 to 50 nm (*Flanagan et al.*, 2006). The uptake of lipophilic components is a process that may be visually observed. If a lipophilic ingredient is dispersed in a surfactant solutions, the solution initially becomes turbid since the dispersed lipophilic ingredient forms large droplets that scatter light. Upon solubilization, the appearance of the solution decreases since the ingredient is incorporated into micelles, which do not scatter light. If sufficient surfactant is available, the solution will eventually become transparent (*Weiss et al.*, 2009).

**Liposomes**

Liposomes or lipid vesicles are structures composed of a bilayer shell that surrounds an aqueous core. They can be manufactured by dispersion of amphiphilic polymers such as polar lipids or mixtures of polar lipids with cholesterol or ergosterol (*Taylor et al.*, 2005; *Laye et al.*, 2008). Due to the molecular geometry of polar lipids, they have the tendency to self-assemble into planar bilayer membrane structures that can be mechanically bend to form shells around an aqueous core. (*Weiss et al.*, 2009) (**Figure 1**). Depending on the way that the polar lipids are dispersed, liposomes may also consist of not just a single but multiple bilayer membranes (*Taylor et al.*, 2005). Liposomes having multiple layers of polar membranes not unlike the shells of an onion are referred to as multilamellar vesicles or MLVs. Liposomes having individual liposomes dispersed in them are referred to multivesicular vesicles, or MVVs (*Gomez-Hens et al.*, 2006). The size of liposomes can vary greatly from as little as 30 nm to as large as 50 μm depending on the characteristics of the polar lipid used (type, concentration), the manufacture conditions (application of shear or ultrasound), as well as environmental conditions (e.g. pH, ionic strength, temperature).

To date, one of the most commonly used polar lipids is phosphatidylcholin (PC). It consist of two nonpolar fatty acid chains and a polar phosphatidyl residue ester linked to the glycerol backbone (*Koynova et al.*, 1998). Commercially, complex mixtures of phospholipids are used, known as lecithins. These may be obtained by a relatively simple extraction process from plant or animal sources such as soy or egg. Because of their chemical structure phospholipids are oriented such that the polar head groups are exposed to both the inner and outer surface of the bilayer, i.e. the two surfaces are both hydrophilic and typically carry a negative charge. As such - despite being very thin - they can serve as very effective barriers to negatively charged species (*Taylor et al.*, 2005). Liposomes are able to encapsulate both lipophilic and hydrophilic components. Hydrophilic ingredients can be contained in the aqueous core while the fatty acid chains of the polar lipids in the bilayer may serve as a lipophilic host (*Weiss et al.*, 2009). The process of accumulating lipophilic ingredients inside the bilayer membrane is referred to as ad-solubilization (*Weiss et al.*, 2009).

A wide variety of manufacturing procedures may be applied to produce liposomes including thin-film rehydration, freeze-drying rehydration, reverse-phase evaporation, detergent depletion, membrane extrusion, high-pressure homogenization, and ultrasonication. An in-depth review of these methods can be found elsewhere (*Taylor et al.*, 2005).

**Biopolymer dispersions and aggregates**

When two biopolymer solutions are combined, a number of different structures may be formed depending on the nature of the biopolymers, the solution composition, and the environmental conditions (*McClements et al.*, 2009). The Flory Huggins theory, which is an extension of the general solution theory, has been used to explain phase separation phenomena that may be encountered when combining similarly charged or uncharged polymers. Restrictions in the freedom of orientation reduce entropy contributions to free energy minimizations thereby favouring separation instead of mixing. As such, the two solutions may never mix and instead form two separate phases being rich in just one polymer (*De Kruif et al.*, 2004). These two phases may in some way be treated like oil and water, and subjected to mild shear stresses, they form dispersions of one in the other. For example, if solutions of pectin and whey protein isolates are combined, allowed to phase separate and then stirred, water-in-water (w/w) "emulsions" can be formed with e.g. whey protein isolate droplet being dispersed in a pectin solution. These systems are very fragile and crosslinking is required to stabilize the whey protein isolate droplets.

On the other hand, if biopolymers of opposite charge are combined, biopolymer aggregates are formed (*De Kruif et al.*, 2004). These complexes may be soluble or insoluble depending on the size and density of the aggregates formed, which depends on the nature of the biopolymers, solution properties and the environmental conditions (*McClements et al.*, 2009). Such aggregates have of late be used to stabilize for example emulsion interfaces (*Salminen et al.*, 2013). The aggregation process itself may also be used to encapsulate components that are incidentally in the presence of the complex former. An example of this is the encapsulation of probiotic bacteria. There, bacteria and two interacting polymers were combined to form a particulate system with probiotic microorganisms being enclosed by a mixed biopolymer matrix. When ingested, the probiotic bacteria were shown to have much higher rates of survival than unencapsulated cells (*Gerez et al.*, 2012).

Interestingly, one can transition from the phase separation scenario to the aggregation one by simply altering pH. If a protein is used, charge reversal occurs at the isoelectric point (p*I*). If the protein is combined with e.g. a negatively charged polysaccharide such as pectin above the pI, phase separation will occur. Once the pH is altered and drops below the pI, the protein becomes oppositely charged to the polysaccharide and electrostatic attraction can take place leading to the formation of aggregates. This is currently an area of active investigation (*Thongkaew et al.*, 2011, *Thongkaew et al.*, 2013).

**Biopolymer-coated emulsions and liposomes**

Using the so called layer-by-layer (LbL) electrostatic depositioning technique, a polyelectrolyte layer may be formed on the surface of a template structure such as an emulsion or a liposome by a charged polymer (*McClements et al.*, 2009). Prerequisite is that the surface of the template structure and the charged polymer carry opposite charges so that electrostatic attraction occurs which drives a directed self-assembly process (*Sukhorukov et al.*, 1998; *Caruso et al.*, 2000). In this way, a single biopolymer monolayer may be build up around the emulsion droplets or the liposomes. The process may be repeated many times to build up multiple layers. The order of mixing, prevailing pH, and ionic strength play a key role in the assembly of such multilayered membranes since they influence the degree of ionization of the charged functional groups on the respective biopolymers (e.g. amino and carboxyl groups) (*Guzey at al.*, 2006). The charge of the biopolymers in turn affects the electrostatic interactions responsible for the attractive forces that lead to the deposition of biopolymers onto the droplets' or liposomes' surfaces and, therefore, assembles into the desired structure (*Schmitt et al.*, 1998). Moreover, charge reversal necessarily occurs after adsorption of charged polyelectrolytes to the surface when the total number of charges is greater than that required to neutralize the oppositely charged template surface (*Guzey et al.*, 2006). This is also a basis for the monolayer formation since the saturated surface repels any excess non-adsorbed polyelectrolyte (**Figure 1**). The charge reversal also allows for adsorption of other oppositively charged polyelectrolyte on top of the first layer (*Guzey et al.*, 2006; *McClements et al.*, 2009).

This technique has first and foremost been developed to functionalize surfaces e.g. by depositing biomolecules on the surface of receptors to act as sensors (*Decher et al.*, 1992; *Decher*, 1997). In the food sector, the technology has been used to increase the stability and functionality of conventional emulsions (*Decher et al.*, 1992; *Guzey et al.*, 2006). To illustrate the process the example of formation of multilayered oil-in-water emulsions may be considered. There, an oil-in-water emulsion is prepared first by homogenizing oil with an aqueous solution in the presence of a charged water-soluble emulsifier resulting in a "primary" emulsion (*McClements et al.*, 2007). The emulsion is then added to a solution containing an oppositively charged biopolymer, e.g. a protein or a polysaccharide at an appropriate concentration. This substrate rapidly adsorbs on the surface of the primary emulsion to form a "secondary" derivative emulsion. This secondary emulsion now consists of oil droplets coated having a surfactant and a biopolymer interfacial layer. Repetition of this procedure allows for oil droplets to be coated with many more biopolymer layers leading to the generation of tertiary and so on emulsions (*McClements et al.*, 2009). The utilization of LbL technique can be used to tailor interfacial membranes in terms of composition, thickness, charge, permeability, and environmental resistance. It is therefore a valuable tool to improve the functionality of oil-in-water emulsions in general (*McClements et al.*, 2009). Previous studies have demonstrated that multilayered emulsions have better stability against environmental stresses than conventional emulsions coated with a single layer (*Gu et al.*, 2004; *Aoki et al.*, 2005; *Surh et al.*, 2005; *Guzey et al.*, 2006). They are able to withstand freezing, drying, heating and shearing processes better than their uncoated counterparts. The LbL approach has also been used to improve the stability and functionality of liposomes (*Henriksen et al.*, 1994; *Were et al.*, 2003; *Laye et al.*, 2008). Using this technique, liposomal integrity during spray drying could be maintained (*Karadag et al.*, 2013). Moreover, this technique has been used to develop colloidal dispersions that allow for a controlled or triggered release of active components or controlled enzyme reactions (*Gu et al.*, 2007).

**Solid lipid nanoparticles**

Solid lipid nanoparticles (SLN) were initially developed in the pharmaceutical industries to deliver lipophilic bioactive compounds. SLN consist of a solid lipid dispersed in an aqueous phase with the bioactives being a part of the lipid matrix (*Westesen et al.*, 1997; *Jenning et al.*, 2000; *Müller et al.*, 2002; *Müller et al.*, 2002; *Schubert et al.*, 2005; *Souto et al.*, 2005; *Souto et al.*, 2006) (**Figure 1**). The particles are stabilized by a surfactant layer, which may consist of a single surfactant, but more often is composed of a mixture of surfactants (*Jenning et al.*, 2000). SLN can be manufactured from either emulsions or microemulsions yielding particles with distinctly different properties. In general, the use of crystallized lipids instead of liquid lipids has shown to improve control over release and stability of incorporated bioactives. This is because mobility of bioactives and interactions with reactive oxygen species can be limited by trapping them in a solid lipid matrix (*Videira et al.*, 2002; *Zhang et al.*, 2004; *Wang et al.*, 2006).

The formation of solid lipid nanoparticles depends on many parameters; chief amongst them the lipid type. Others include surfactant type, surfactant concentration, droplet size, and cooling conditions (*Muller et al.*, 2004; *Wissing et al.*, 2004; *Lombardi Borgia et al.*, 2005). The choice of the lipid carrier generally dictates the conditions under which the SLN has to be manufactured, e.g., homogenization temperature and cooling rates, since the lipid-bioactive mixture first has to be melt-homogenized and then solidified by cooling. The chosen lipid will also affect the achievable loading capacity for a given lipophilic bioactive since the bioactive has to be solubilized in the melt prior to homogenization in order to form a mixed bioactive-lipid core. Finally, the lipid type also affects the release characteristics of the encapsulated bioactive from the SLN, which depends on melting and lipolytic degradation characteristics. These prerequisites are in many cases best fulfilled by triacylglycerides (TAGs), which are therefore most commonly used as carrier lipids.

TAGs however exhibit polymorphism upon cooling; that is, the individual chains of the lipid molecules may assume a variety of possible configurations giving rise to longitudinal stacking of TAG molecules in lamellae that lead to the formation of α, β′, and β crystals with hexagonal, orthorhombic, and triclinic unit structures, respectively. These crystals differ in their thermodynamic stability - a fact that is expressed in the different melting points of the crystals, with α being the most unstable and β being the most stable form of a given TAG. As such, SLN containing high amount of the α-polymorph tend to recrystallize in the β-form over time. On the other hand, due to differences in the crystal morphologies, SLNs having high amount of the α-polymorph tend to be spherical, while SLN with high amount of the β-polymorph form platelets or needles (*Mehnert et al.*, 2001). The crystal morphology of the α-polymorph appears to also facilitate a higher loading with bioactives, possibly due to a larger lattice spacing between TAGs.

Recent research has shown that the surfactant plays a key role in controlling the crystallization process. Because of the small size of the parent emulsion, the number of emulsifier tails that directly interact with the TAGs is large enough to modulate the crystallization process – a phenomena also referred to as surface-initialized crystallization. Moreover, the surfactant has a substantial impact on the subsequent kinetic stability of the generated crystal structure since it can provide a certain thermodynamic resistance against a recrystallization. As such, choosing the right surfactant allows one to create SLNs with high amount of the α-polymorph even if that crystal structure is thermodynamically less stable than that of the corresponding alternative polymorphic form (*Bunjes et al.*, 2003). Finally, the surfactant choice is crucial to stabilizing the melt-homogenate that serves as a pre-cursor for the SLNs. Fulfilling these three key demands with a single surfactant is difficult if not impossible and mixtures of nonionic and ionic surfactants have therefore shown to be ideal choices to generate highly functional SLN (*Wörle et al.*, 2006).

Particle size of the SLN is a further parameter that affects stability. Albeit most SLN are made with sizes ranging between 30 – 200 nm, smaller SLN may have different stabilities and (biological) functionalities than larger ones (*Westesen et al.*, 1993; *Jenning et al.*, 2000). A reduction in particle size leads to increased surface curvatures which may allow incorporated bioactives to more easily be accessed and liberated. Moreover, the amount of α-polymorph may increase the smaller the particle since the relative amount of TAGs interacting with surfactant increases leading to altered thermal stabilities. Conversely, if the particle size increases, e.g. because of partial coalescence, the system begins to behave more like a bulk fat crystallizing at higher temperatures (*Siekmann et al.*, 1994; *Jenning et al.*, 2000; *Palanuwech et al.*, 2003).

Aside from the surfactant and lipid choice, process conditions - and here especially the cooling rate – affects crystal formation. In appropriate cooling speeds can lead to the formation of plated-shaped particles that quickly aggregate to form a gel.. Figure 4 shows results of a simple melt experiment after having cooled a nanoemulsion composed of tripalmitin and stabilized by Tween 20 at 0.2, 2, and 20 °C/min (*Helgason et al.*, 2008). The differential scanning calorimetry thermogram reveals that as the cooling speed increases from 0.2 to 20 °C/min, the concentration of α-crystals (visible by the transition peak at ~43 °C) increases while the concentration of β-crystals that are formed at ~61 °C in the SLN suspension remains virtually unchanged. The effect of cooling speed on crystal formation has also been reported by *Jenning et al.* (2000) who found that cooling at 10 °C/min resulted in formation of α-crystals while at 2 °C/min, β-crystals were predominately formed.

**Colloidosomes**

Colloidosomes are particles with a core-shell structure that may be used as microcapsules and to control release of active compounds in pharmaceuticals, foods, and cosmetics (*Kim et al.*, 2007; *Rossier-Miranda et al.*, 2009). The external shell is composed of closely packed uniform colloidal particles (*Dinsmore et al.*, 2002; *Kim et al.*, 2007; *Rossier-Miranda et al.*, 2009). The nature of the core depends on the template structure that serves as acceptor for the colloidal particles. Suitable template structures are emulsion droplets, solid particles, gas bubbles etc. The formation of colloidosomes may be achieved by either (i) electrostatic depositioning or (ii) adsorption of small colloidal particles on the surface of larger ones (*Gu et al.*, 2007; *Jiang et al.*, 2008). This first is based on the LbL electrostatic deposition method that has been described above to prepare biopolymer-coated food dispersions (*Guzey et al.*, 2006). There, a base dispersion is produced using a charged surfactant. The dispersed system is then added to a solution containing oppositely charged solid particles (e.g. solid lipid nanoparticles or biopolymer aggregates) causing a shell of particles to form. The production of colloidosomes by electrostatic attraction allows for a flexible engineering of interfaces in terms of particle size, surface charge, and surface density (*Gu et al.*, 2007). The second approach involves vigorous mixing or homogenization of small colloidal particles such as polystyrene or latex spheres with the two major constituents of the dispersion (e.g. oil and water) in the absence of a surfactant. If the surface energies of the spheres and the interfacial tension between the two dispersion constituents has appropriate values, the colloidal particles tend to adsorb to the interface that are generated during the mixing or homogenization (*Dinsmore et al.*, 2002). Emulsion interfaces stabilized by solid particles are also known as "Pickering" emulsions. The technique has also been used to for example generate "hairy" particles (*Noble et al.*, 2004). There, the investigators used polymeric rods to stabilize a water-in-oil emulsion. This was followed by gelation of the water phase after which the oil phase was removed

Colloidosomes may greatly vary in their characteristic dimensions, which may be attributed to the large amount of food dispersions that can be used to generate them (*Rossier-Miranda et al.*, 2009). Colloidosomes have for example been created from water-in-oil emulsions, oil-in-water emulsions, and water-in-oil-in-water emulsions based colloidosomes (*Saraf et al.*; *Hsu et al.*, 2005; *Gu et al.*, 2007). The formation of colloidosomes provides an efficient mean to control size, permeability, mechanical strength and compatibility of a dispersions (*Simovic et al.*, 2008). The membrane pore size of the shell can be engineered by varying the size of the small solid particles of which the shell consists, and by controlling their degree of fusion (*Saraf et al.*; *Kim et al.*, 2007). If heat treated, the particles may be sintered together, creating bridges or "necks" between them. Such shells are stable and rigid. Alternatively, the interstitial pores may be filled with smaller particles or polymers (*Dinsmore et al.*, 2002). Adsorbed particles can also be linked together by electrostatically depositing additional polyelectrolytes onto the shell (*Dinsmore et al.*, 2002; *Gu et al.*, 2007; *Simovic et al.*, 2008). As stated above, the core material may also be removed by dissolution leaving just the shell behind (*Dinsmore et al.*, 2002; *Noble et al.*, 2004).

**Nanofibers**

The production of fibers with diameters of less than 100 nm has become feasible with the discovery of the electrospinning process (*Zhang et al.*, 2005; *Kriegel et al.*, 2008). Electrospinning is a manufacturing technology capable of producing thin, solid polymer strands from solution by applying a strong electric field to a spinneret equipped with a small capillary orifice. Generally, electrospun polymer fibers can range in diameters from 10 to 1000 nm. Due to the high surface to volume ratio they may exhibit unusual functionalities with respect to their mechanical, electrical, and thermal behaviour. They have been used to produce tissue templates, medical prostheses, protective clothing, and electronic devices. Various studies have explored their use in the food industry (*Kriegel et al.*, 2008).

The principle of electrospinning involves the formation of a small drop of polymer solution at the tip of a syringe. The syringe is connected to a high voltage generator with voltages typically exceeding 20 kV. Opposite from the syringe a grounded collector plate is mounted. The strong electrostatic field causes the polymers in the solution to be so strongly charged that they distort the droplet shape to form a so-called Taylor cone. At the tip of the Talor cone, a very thin charged polymers solution jet is expelled, which is accelerated towards the ground target. Charge instabilities cause the polymer jet to bend and stretch during its travel to the collector which causes to the jet to become even thinner. On the way to the collector plate, the solvent evaporates and the jet becomes solid to form a single fiber strand. This fiber strand forms a film on the surface of the collector plate that can later be collected for use. If the polymer concentration is not high enough or the polymers are not entangled, small droplets (or beads) rather than a fiber are deposited on the collector plate. At intermediate concentrations, one may obtain a mixture of fibers and beads. Whether beads or fibers are obtained also depends on the applied voltage, i.e. at smaller voltages droplets are electrosprayed while at higher voltages fibers are electrospun.

The majority of studies with nanofibers has focused on the production and reinforcement of conducting nanotubes for use in next-generation microprocessors. Interestingly though, most patents in this subject area describe uses in life sciences. For example, biocompatible nanofibers have yielded porous membranes for skin to aid in cleansing, healing, and dressing of wounds. Tubular fibers for blood vessel and nerve regeneration, three-dimensional scaffolds for bone and cartilage regeneration, and drug-delivery matrices have been produced by electrospinning.

In all these cases, the unusually high difference between the two primary dimensions thickness and length has been exploited. Due to the length of the fiber, the system forms a solid film that can be handled by conventional textile technologies. The small thickness though facilitates an extremely rapid dissolution, which is highly desirable to stop a heavily bleeding wound b inducing clotting. Such fibers have also shown to serve as ideal templates for cell deposition and proliferation. The exact reason why cells grow extremely well on such fiber mat surfaces is not yet fully clear. Beyond biomedical applications, electrospinning has been applied to produce filter media for liquid and gas filtration and protective clothing for the military that is capable of trapping aerosols and large molecular-weight biocidal gases while minimally impeding air flow. Thermal, piezoelectric, and biochemical sensors made from electrospun fibers have demonstrated sensitivities that are 2 or 3 orders of magnitude higher than comparable thin films.

The food industry has only recently begun to explore the use of electrospun. It has been postulated that they could be used as novel food packaging material, as structiural elements of a food matrix to provide for texture, and as a scaffolding for bacterial cultures (*Ignatova et al.*, 2007). While the number of applications that make use of electrospun fibers is increasing at an exponential rate, the applications for food and agricultural systems to date remain few. This is because it has only recently been understood how to produce fibers from biopolymers using water as a solvent. Most studies utilized instead well-defined synthetic polymers dissolved in an organic phase, an approach that could not be used in food applications. However, recent studies have reported significant progress in electrospinning a wide variety of biopolymers in water or weak acids, which will likely increase the number of food applications.

## STRUCTURE-AFFECTING ENZYMES

The use of enzymes has been well established in the food industry, often with the purpose of improving the quality and shelf life of foods or to create modified ingredients. The specific control of enzymatic reactions and the discovery and exploitation of new and to date unknown enzymes require a thorough understanding of the biophysical basis of the mode of action as well as the catalyzed chemical reaction (*Kennedy et al.*, 1996). While aroma creation, color stabilization, bacterial growth inhibition, and reactive oxygen species depletion may often be a reason to use enzymes in foods, there is a yet not fully exploited potential to engineer food structures. One may simply think of the case of beer or wine clarification by enzymatic means. There, enzymes cause an aggregation and sedimentation of colloidal biopolymer aggregates that scatter light. As a result the previously cloudy beverage becomes transparent. While this is only due to a very minor chemical change in terms of composition, the effect on structure is quite profound (*Minussi et al.*, 2002).

*Gerrard et al.* (2002) not too long ago provided a very good overview over the use of enzymes to modify the functional properties of food. The advantage of using enzymes compared to other methods such as for example chemicals are that reactions conditions are much milder, transformations are highly specific, and byproduct generation is lower leading to the generation of less toxic compounds (*Gerrard*, 2002; *Gübitz et al.*, 2003; *Rodríguez Couto et al.*, 2006). Of particular significance for texture and structure modification are enzyme-catalyzed reactions that crosslink proteins and/or polysaccharides. The enzymes catalyze a reaction in which inter- and intramolecular bonds are formed in and between biopolymers. This can lead to modification of their solubility, foaming, whipping and emulsifying properties (*Matheis et al.*, 1984; *Matheis et al.*, 1987).

Different types of crosslinks are prevalent in food products. One of the most commonly occurring covalent crosslink is the disulfide (S-S) crosslink formed by oxidative coupling of two cysteine residues (*Dickinson*, 1997; *Gerrard*, 2002). The ability of proteins to form these disulfide bonds is often the basis of network formation and gelation. Another interesting crosslink is that of the isopeptide bonds that can be formed between protein residues. Aside from the two above mentioned cases, many other types of crosslinks not only with proteins but also with polysaccharides are possible. Food technologists have only recently begun to truly explore the opportunity to generate novel gel-like network structures and emulsified food systems. As an introduction to this subject, enzyme-catalyzed crosslinks and the involved substrate polymers are shown in **Table 1**. Below, an overview of the most important enzyme systems, their reaction mechanism, and their physiological roles will be given.

**Transglutaminase (EC 2.3.2.13)**

Transglutaminase belongs to the class of (acyl-) transferases. Enzymes belonging to this class catalyze an acyl-transfer reaction between the γ-carboxyamide group of a peptide-bound glutamine residues and a variety of primary amines (*Folk et al.* 1977, 1985; *Greenberg et al.*, 1991; *Motoki et al.*, 1998). The γ-carboxyamide group of a peptide-bound glutamine residues acts as acyl donor, whereas the primary amines act as acyl acceptor. When an ε-amino group of a lysine residue bound in a peptide serves as substrate, two peptide chains become subsequently covalently linked (*Folk et al.*, 1977). A new isopeptide bond called ε-(γ-glutamyl)lysine bond is formed and ammonia is released (*Tanimoto et al.*, 2002). Structural conformation of proteins influence the accessibility to its glutamine and lysine residues, and for this reason the ability of the enzyme to crosslink the proteins (*Kellerby et al.*, 2006). It should be noted though that two other reactions are also catalyzed by transglutaminase (**Figure 2**). In the absence of amines, transglutaminase is capable of hydrolyzing the γ-carboxyamide group of the glutaminyl residues causing a deamidation of glutamine (*Dickinson*, 1997). Here, water molecules act as acyl acceptor. Transglutaminase can also incorporate amine residues in proteins (*De Jong et al.*, 2002).

Transglutaminase is a ubiquitous occurring enzyme which is widely found in nature. For example, it is present in several animal tissues and their organs, in marine and in plant life (*Folk et al.*, 1985). Transglutaminase is involved in different important biological actions. It plays for example a key role in wound healing and blood clotting. The enzyme was initially isolated form guinea pig liver, but this proved to be an impractical source for commercialization because of the high costs of extraction and purification. Moreover, this transglutaminase was calcium ion dependent, which is a somewhat inconvenient dependency because various proteins, such as caseins, soy globulins and myosins are also sensitive to $Ca^{2+}$. The addition of calcium would have thus simultaneously led to undesirable changes in protein functionality. Today, microbial transglutaminase is used instead. This type of transglutaminase derived from a variant of *Streptoverticillium mobarense* is extracellular and secreted into the culture medium. This allows for an easy separation and purification of the enzyme since cell disruption can be avoided. Moreover, the commercially generate microbial transglutaminase is calcium independent and therefore widely applicable (*Zhu et al.*, 1995; *Motoki et al.*, 1998).

**Table 1 Crosslinking enzymes and their substrates; adapted and modified from *Lantto et al.* (2007).**

| enzyme | reaction type | target substrate | note | reference |
|---|---|---|---|---|
| Hydroxyl oxidase (EC 1.1.3.X) | Acting on CH-OH group of donor with $O_2$ as acceptor | OH-groups | | |
| Glucose oxidase (EC 1.1.3.4) | Oxidation of glucose to form glucono-δ-lactone | Glucose | | (*Lantto*, 2007) |
| Lysyl oxidase (EC 1.4.3.13) | Oxidative deamidation of lysine | Lysine | Intermediates act with aldyhydes and form linkages | (*Matheis et al.*, 1987) |
| Sulfhydryl/Thiol oxidase (EC 1.8.3.X) | Oxidation of cysteine residues to form disulfide bonds | Cysteine | | (*Aurbach et al.*, 1962), (*Thorpe et al.*, 2002) |
| Proteindisulfid reductase (EC 1.8.4.2) | Reaction between gluthathione and protein-disulfide | Glutathione | | |
| Peroxidase (EC 1.11.1.7) | Radical-generating oxidation of aromatic compounds | Tyrosine | Acts with carbohydrates containing ferulic acid, caffeic acid or chlorogenic acid residues | (*Matheis et al.*, 1984), (*Veitch*, 2004) |
| Lipoxygenase (EC 1.12.11.12) | Radical-generating oxidation of unsaturated fatty acids | Polyunsaturated fatty acids | $H_2O_2$ is capable of forming linkages | (*Matheis et al.*, 1987) |
| Polyphenol oxidase (EC 1.14.18.1) | Radical-generating oxidation of aromatic compounds | | | (*Matheis et al.*, 1987), (*Matheis et al.*, 1984) |
| Transglutaminase (EC 2.3.2.13) | Formation of isopeptide bond | Protein-bound glutamine & lysine | | (*Motoki et al.*, 1998), (*Folk et al.*, 1977) |
| Proteindisulfid isomerase (EC 5.3.4.1) | Rearrangement of S-S bonds in proteins | Cysteine | | (*Wilkinson et al.*, 2004) |

Microbial transglutaminase exhibits enzymatic activity over a wide pH range. The pH optimum is around 5 to 8, and the temperature optimum is at 50 °C. The enzyme loses its activity upon heating to 70 °C within a few minutes (*Motoki et al.*, 1998; *Gerrard*, 2002; *Ajinomoto*, 2009).

A  Gln-CO-$NH_2$ + $NH_2$-Lys ⟶ Gln-CO-NH-Lys + $NH_3$

B  Gln-CO-$NH_2$ + $NH_2$-R ⟶ Gln-CO-NH-R + $NH_3$

C  Gln-CO-$NH_2$ + $H_2O$ ⟶ Gln-CO-OH + $NH_3$

**Figure 2 Different reactions catalyzed by the enzyme transglutaminase (*Motoki et al.*, 1998); (A) acyl transfer reaction, (B) crosslinking reaction between Gln und Lys residues of proteins and peptides; (C) deamidation.**

### Polyphenol oxidase (EC 1.14.18.1)

The term polyphenol oxidase refers to enzymes oxidizing phenols including tyrosinase and laccase with the concomitant reduction of oxygen to water. They are found in many plants and are often also referred to phenolases, phenol oxidases, and catechol oxidases; depending on their source (*Rescigno et al.*, 2002). Polyphenol oxidases contain a di-copper center acting as an oxygen carrier. Polyphenol oxidases form quinones that are highly reactive species and powerful electrophiles, able to initiate non-enzymatic reactions following different pathways to produce polymers such as lignin (*Claus et al.*, 2006; *Octavio et al.*, 2006). These enzymes are also involved in enzymatic browning which is one of the most important color reaction occurring in plants, fruits, and vegetables. Below, the enzymes laccase and tyrosinase are described in more detail.

#### *Laccase (EC 1.10.3.2)*

Laccase is a polyphenol oxidase and thus an oxidoreductase. It oxidizes polyphenols, methoxy-substituted phenols, anilines, diamines and a substantial range of other intermediates (*Minussi et al.*, 2002). This enzyme has an overlapping substrate range with other type of oxidases especially tyrosinase, but is not capable of oxidizing tyrosine like tyrosinase. Laccase oxidizes *p*-diphenols in the presence of oxygen, unlike tyrosinase, which is only able to oxidise *o*-diphenol and to hydroxylate monophenols (*Carunchio et al.*, 2001). The substrate oxidation by laccase is a one-electron reaction. Laccase has multiple copper centers transferring electrons from the reducing substrate to molecular oxygen. This reaction mechanism produces non-toxic peroxides. Subsequently, the electrons are passed through the trinuclear cluster of the reactive site, where molecular oxygen is reduced and water is released. The oxidation of the substrate generates typically unstable reactive radicals which undergo non-enzymatic reactions (*Thurston*, 1994; *Claus*, 2004). As a result of the radical production, a crosslinking of monomers, as well as a degradation of polymers, and cleavage of rings is possible. Of particular interest to food scientists is the crosslinking of monomers. The oxidation of phenols or anilines by laccases generates radicals which further react with each other to form biopolymers that are covalently connected by C-C, C-O, and C-N bonds (*Claus*, 2004). A typical laccase-catalyzed transformation is illustrated in **Figure 3**. *Bunzel et al.* (2001, 2004, 2005) identified a whole spectrum of dehydrodiferulic and dehydrotriferulic acids in cell walls of cereal fibers such as maize, wheat, and rice that may serve as substrates. Furthermore, in the presence of a mediator molecule, laccase is capable of oxidizing nonphenolic compounds which cannot be oxidized by the enzyme itself (*Bourbonnais et al.*, 1990; *Niku-Paavola et al.*, 2000). The optimal pH range for most laccases is around 4.0 to 5.0 (*Mattinen et al.*, 2006).

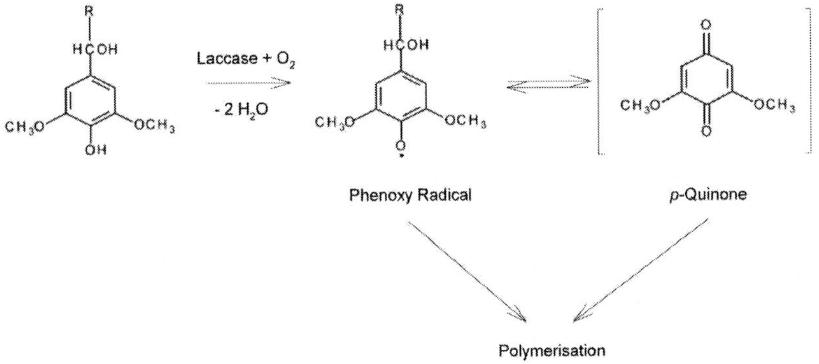

Laccase + $O_2$

$-2\,H_2O$

Phenoxy Radical

*p*-Quinone

Polymerisation

**Figure 3 Laccase-catalyzed oxidation of phenols (*Minussi et al.*, 2002).**

This multinuclear copper enzyme is ubiquitously occurring (*Claus*, 2004). It belongs to a small group of enzymes called the large blue copper proteins or blue copper oxidases (*Thurston*, 1994). Laccase can be obtained from bacteria, fungi, and plants such as *Agaricus bisporus*, *Podospora anserina*, and *Rhizoctonia practicola* to name just a few (*Yaropolov et al.*, 1994; *Gianfreda et al.*, 1999; *Mayer et al.*, 2002). In fungi or higher plants laccase is responsible for the formation of pigments such as melanin and for the degradation of lignin (*Octavio et al.*, 2006).

*Tyrosinase (EC 1.14.18.1)*

Tyrosinase is an oxidoreductase which converts phenols into *o*-quinones that can then react with several nucleophiles (*Aberg et al.*, 2004). The enzyme is bifunctional and catalyzes two different reactions: the *o*-hydroxylation of monophenols and later the oxidation of *o*-diphenols to *o*-quinones, both using molecular oxygen (*Sánchez-Ferrer et al.*, 1995; *Selinheimo et al.*, 2006). These two activities are termed as cresolase or monophenolase and catecholase or diphenolase activities, respectively, as shown in **Figure 4**.

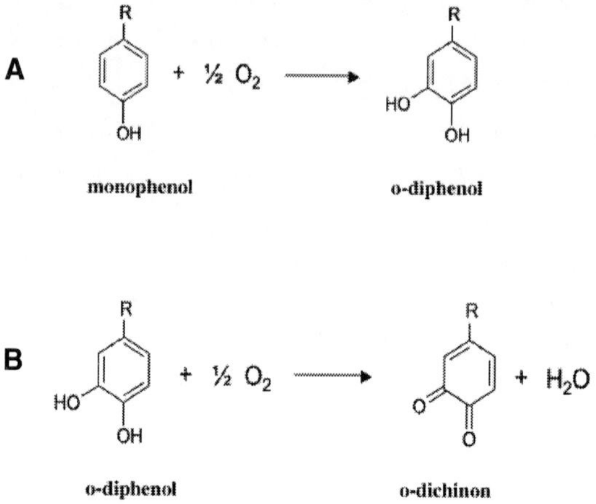

**Figure 4 Enzymatic activities of tyrosinase comprising both reactions, (A) monophenolase (cresolase) and (B) diphenolase (catecholase) (*Claus et al.*, 2006).**

There, *o*-monophenol is oxidized to create an *o*-diphenol with one oxygen molecule being reduced to water, and the other one being incorporated into the phenolic ring. The second step is a further oxidation of the *o*-diphenol. As a result of this oxidation one obtains a corresponding *o*-quinone. These quinones are highly reactive components and can spontaneously polymerize to form macromolecular biopolymers (*Claus et al.*, 2006; *Lantto*, 2007). Oxidation of tyrosine residues in proteins leads to the corresponding quinones that can then react with e.g. free sulfhydryl (thiol) and/or amino groups. Result of this coupling is the formation of tyrosine-cysteine and tyrosine-lysine linkages. Quinones are also able to crosslink and to form tyrosine-tyrosine bonds. After step-wise oxidation of phenolic molecules such as caffeic acid or chlorogenic acid to *o*-quinones the addition of a protein leads to a coupling (**Figure 5A**). It is also possible to enzymatically couple tyrosine residues of different proteins resulting in a direct protein-protein cross link as illustrated in **Figure 5B** (*Jus et al.*, 2008).

Tyrosinase is a copper-containing enzyme and widely found in nature. Tyrosinase is for example present in prokaryotic and eukaryotic microorganisms, as well as in mammals, invertebrates and plants. It is essential for the formation of pigments such as melanin (enzymatic browning of food) and is important in wound healing and primary immune response (*Thalmann et al.*, 2002). The tyrosinase gene is derived from *Streptomyces castaneoglobisporus* (*Lantto*, 2007).

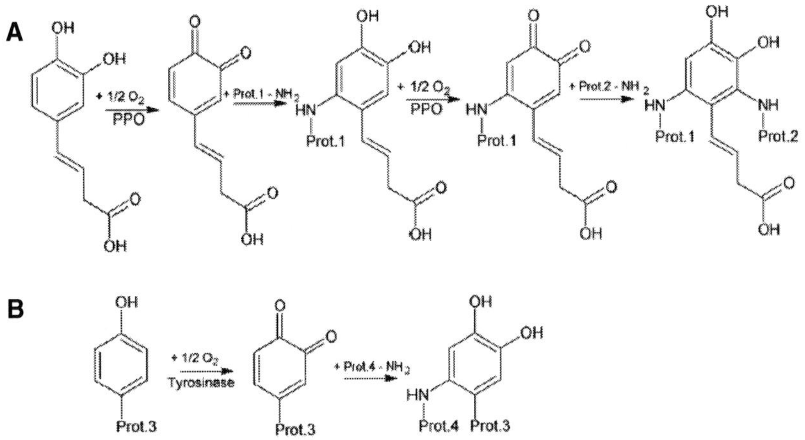

**Figure 5 Reaction principle of tyrosinase-catalyzed crosslinks: (A) in the presence of caffeic acid, a low molecular weight phenolic reagent; (B) in the absence of a low molecular weight phenolic reagent (*Thalmann et al.*, 2002).**

### Peroxidase (EC 1.11.1.7)

Peroxidase utilizes hydrogen peroxide to oxidize a wide range of organic and inorganic compounds. The following reaction is characteristic for peroxidase (*Veitch*, 2004):

$$H_2O_2 + 2AH_2 \xrightarrow{\text{HRP C}} 2H_2O + 2AH^{\bullet}$$

$AH_2$ and AH* are the reducing substrate and its radical product, respectively. Aromatic phenols, indoles, amines, and sulfonates are typical substrates. Peroxidase is able to catalyze different reactions depending on the substrate: peroxidation, oxidation, catalayse, and hydroxylation all shown in **Figure 6** (*Whitaker et al.*, 1984).

Phenols or amines can act as hydrogen donors that facilitate peroxidatic reactions that depend on hydrogen peroxide (**Figure 6A**). As a result quinones are produced. In the absence of any hydrogen donor, peroxidase converts hydrogen peroxide to water and oxygen. This type of reaction is called catalatic transformation (**Figure 6C**). In the presence of molecular oxygen and certain hydrogen donors such as tyrosine, dihydroxyfumaric acid or ascorbic acid peroxidase-catalyzed oxidation (**Figure 6A**) or hydroxylation (**Figure 6D**) is possible (*Whitaker et al.*, 1984). Especially the formed quinones can undergo non-enzymatic polymerization reactions with a variety of food components containing amino and sulfhydryl residues. This may facilitate bond formations such as diferulate linkages of polysaccharids or pectins and the formation of dityrosine links (*Veitch*, 2004). For more information, the reader is referred to a detailed reaction scheme shown in the review of *Whitaker et al.*, (1984).

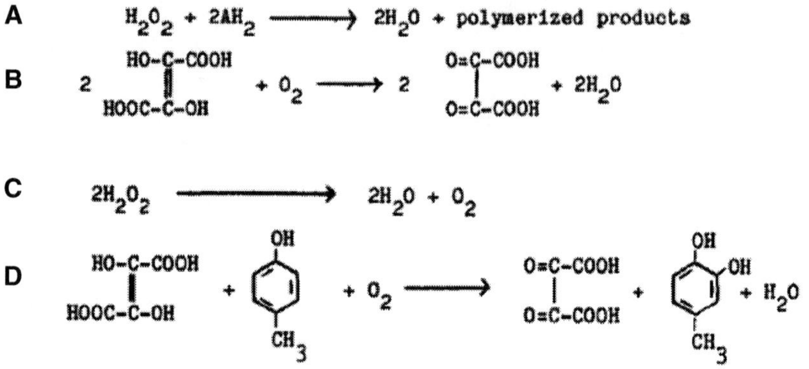

A $H_2O_2 + 2AH_2 \longrightarrow 2H_2O$ + polymerized products

B 2 HO–C–COOH ‖ HOOC–C–OH + $O_2 \longrightarrow$ 2 O=C–COOH | O=C–COOH + $2H_2O$

C $2H_2O_2 \longrightarrow 2H_2O + O_2$

D HO–C–COOH ‖ HOOC–C–OH + (OH, $CH_3$ ring) + $O_2 \longrightarrow$ O=C–COOH | O=C–COOH + (OH, OH, $CH_3$ ring) + $H_2O$

**Figure 6 Peroxidase-catalyzed reactions (*Whitaker et al.*, 1984).**

Peroxidase is also widely found in plants, microbes and animals. A lot of peroxidases are heme-proteins containing a ferric protoheme group. Others have magnesium, vanadium or selenium, or a flavin group at their active site (*Smith et al.*, 1998). It seems, that each plant species contains a group of peroxidase isoenzymes with the potential to carry out a variety of different reactions (*Veitch*, 2004). In plants, peroxidase is responsible for the biosynthesis and degradation of cell wall lignin, is part of the cell's defense against pathogens, and restricts cell growth (*Yuan et al.*, 2003). Horseradish peroxidase is one of the most studied enzymes of all peroxidases (*Lantto*, 2007).

**Sulfhydryl oxidase (EC 1.8.3. )**

Sulfhydryl oxidases (EC 1.8.3) include thiol oxidases (EC 1.8.3.2) and glutathione oxidases (EC 1.8.3.3), which are both capable of catalyzing the formation of disulfide bridges between two thiols (*Lantto*, 2007). Glutathione oxidase reduces $O_2$ to $H_2O_2$ (*Kusakabe et al.*, 1982; *Sliwkowski et al.*, 1984) while in contrast thiol oxidase reduces $O_2$ to water (*Aurbach et al.*, 1962). Both reactions are shown in **Figure 7**.

*Lantto et al.* (2007) reported that the term sulfhydryl oxidase is in fact confusing and classification of sulfhydryl oxidases in the two above-mentioned classes difficult. Both glutathione and thiol oxidase have differences in their substrate specificity. Glutathione oxidase has been isolated from different sources such as rat seminal vesicles, fungi e.g.

**A** glutathione oxidase

$$2\,RSH + O_2 \rightarrow RS\text{-}SR + H_2O_2$$

**B** thiol oxidase

$$2\,RSH + \tfrac{1}{2}\,O_2 \rightarrow RS\text{-}SR + H_2O$$

**Figure 7 Typical reactions of glutathione oxidase (A) and thiol oxidase (B) (*Lantto,* 2007).**

*Penicillium*, *Aspergillus niger* and *Saccharomyces*, and chicken egg white (*Kusakabe et al.*, 1982; *De La Motte et al.*, 1987; *Hoober et al.*, 1996; *Ostrowski et al.*, 2002). Its reactive site contains flavin-depending proteins catalyzing oxidation of small thiol compounds such as glutathione, cysteine or dithiotreitol (*Lantto*, 2007). Thiol oxidase belongs to the metallo proteins, has no flavin-depending group, and is able to catalyze a wider range of substrates. It has been found in bovine milk and other mammalian tissues but not in fungal ones (*Janolino et al.*, 1975; *Blakistone et al.*, 1986). Sulfhydryl oxidase found in bovine milk oxidizes thiol groups from cysteine and cysteine residues from peptides and proteins to induce polymerization reactions via disulfide linkages. The enzyme can however not oxidize the small thiol compound dithiotreitol (*Janolino et al.*, 1975).

**Lysyl oxidase (EC 1.4.3.13)**

Lysyl oxidase is also a member of the class of oxidases. It catalyzes the oxidative deamidation of lysine or δ-hydroxy-lysine residues of e.g. collagen and elastin. As a result of this oxidation α-amino adipic acid δ-semialdehyde (allysine) and δ-hydroxy α-amino adipic δ-semialdehyde (hydroxyallysine) are generated. These intermediates are able to react with other aldehydes or with lysine, δ-hydroxylysine and histidine residues to form intra- and intermolecular cross links. The use of lysyl oxidase as a cross linking agent in food products has not yet been studied (*Matheis et al.*, 1987). A reason for this may be that lysyl oxidase is not commercially available.

**Glucose oxidase (EC 1.1.3.4)**

Another member of the class of oxidases is glucose oxidase which is capable of oxidizing β-D-glucose to gluconic acid and hydrogen peroxide, as illustrated below:

$$C_6H_{12}O_6 + H_2O + O_2 \rightarrow C_6H_{12}O_7 + H_2O_2$$

Hydrogen peroxide produced during the glucose oxidase reaction induces cross link between free sulfhydryl residues of proteins. Hydrogen peroxide can also form ditryosine crosslinks between proteins residues (*Bonet et al.*, 2006). Glucose oxidases are naturally occurring in fungi such as *Aspergillus* (*Kona et al.*, 2001) and *Penicillium* (*van Dijken et al.*, 1980). It is expressed in the presence of glucose and first oxidizes glucose to form glucono-δ-lactoen and eventually gluconic acid (*Lantto*, 2007).

**Lipoxygenase (EC 1.13.11.12)**

Lipoxygenase uses as substrate polyunsaturated fatty acids. In the presence of molecular oxygen the enzyme is able to form free fatty acid peroxy radicals. These peroxy radicals abstract a hydrogen molecule from other unsaturated fatty acid molecules or from other similar structured components. This process is repeated until no more substrate is available. Hydroperoxides decompose into acids, ketones, aldehydes such as malondialdehyde (*Matheis et al.*, 1987). These intermediate products are highly reactive and form chemical crosslinks with proteins. Potential reactions that can be occur are (*Matheis et al.*, 1984):

(a) Oxidation of cysteine, cysteine, methionine, tryptophan, and histidine side chains by hydroperoxides
(b) Oxidation of sulfhydryl groups of cysteine residues to from disulfide bonds, causing intra- and intermolecular crosslinking of proteins
(c) Protein-protein covalent linkages via free hydroperoxide radicals
(d) Covalent attachment of fatty acids to proteins
(e) Crosslinking of proteins with malondialdehyde

**Figure 8** show a simplified schematic of the reaction catalyzed by lipoxygenase. A detailed description of the reaction scheme can be found in the publication of *Matheis et al.* (1987). In the presence of lipids such as oil-in-water emulsions lipoxygenase-catalyzed reactions often generate undesirable off-flavors. *McClements et al.* (2000) stated that the formation of hydroperoxides is an essential mean to retard lipid oxidation in emulsions.

## USE OF ENZYMES TO ENGINEER FOOD DISPERSIONS

Below, commercially available enzymes, e.g. transglutaminase, laccase, tyrosinase, and glucose oxidase that have substantial potential to improve the stability and functionality of food dispersions are briefly discussed. Special emphasis is given to their application in food systems.

**Transglutaminase**

Transglutaminase (TGase) is able to modify the functionality of proteins by inducing intra- and intermolecular crosslinks (*Nio et al.*, 1986). As discussed above, the enzyme catalyzes acyl-transfer transformations between glutamine and lysine residues in proteins thereby generating isopeptide bonds (*De Jong et al.*, 2002). TGase is a commercially available crosslinking agent whose use is approved for foods. *Dickinson* (1997) for example reported that TGase is able to modify proteins such that stability of protein-stabilized food dispersions is improved, e.g. by facilitating the formation of emulsion droplet networks and by stabilizing protein-coated interfaces. Casein and casein micelles offer excellent emulsifying properties and are also good substrates for TGase. Its crosslinking rate depends on the accessibility of the lysine and glutaminyl residues in the micellar substrate (*Dickinson et al.*, 1996). Casein micelles have relatively flexible structures and are therefore quite suitable substrates for TGase - in contrast to globular whey proteins (*Nio et al.*, 1986; *Aboumahmoud et al.*, 1990). Formation of TGase-induced crosslinks in casein micelles improves their stability when subjected to different treatments such as acidification, heating, and cooling (*Schorsch et al.*, 2000; *Flanagan et al.*, 2003; *O'Connell et al.*, 2003). *Smiddy et al.* (2006) also observed a greater stability of TGase-treated casein micelles against superimposed mechanical forces.

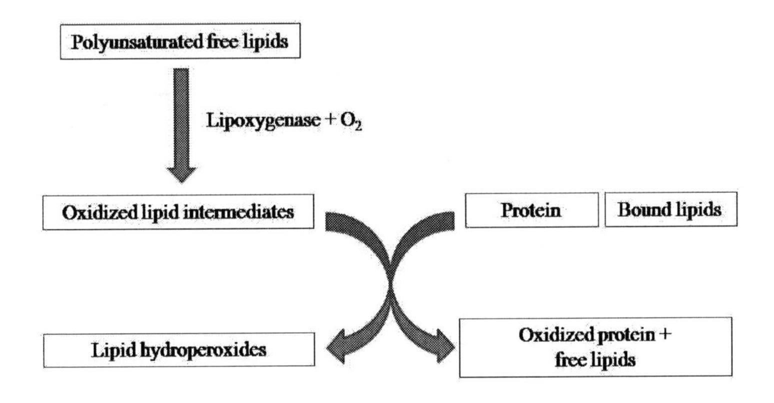


**Figure 8 Simplified scheme of the reaction mechanism of lipoxygenase in dough mixing (*Matheis et al.*, 1987).**

Crosslinking of caseins by TGase leads to the formation of a covalent network that cannot be disrupted by the addition of urea or trisodium citrate. The destabilizing action of citrate is based on its ability to chelate micellar calcium phosphate - a key constituent contributing to the maintenance of the structure of the micelle; thereby affecting micellar integrity (*Huppertz et al.*, 2007). TGase has shown to be able to improve the longterm stability of not just casein micelles but also of caseinate gels. *Dickinson* (2012) for example showed that enzymatic crosslinking produces a rigid protein gel with comparable or superior properties to those of a heat-set gel. TGase treatment of acidified sodium caseinate dispersions led to formation of firm and homogeneous gel structures. (*Myllärinen et al.*, 2007). Another study examined the effect of TGase on milk fat globules and the emulsifying properties of milk proteins. The authors demonstrated that the extent of enzymatic crosslinking of caseins increased with incubation time, and that fat droplets were less prone to coalesce (*Hinz et al.*, 2007). The effect of enzyme-catalyzed crosslinking on interfacial rheology and foam stability was also investigated (*Partanen et al.*, 2009).

Other authors investigated the effect of TGase-catalyzed crosslinking on emulsion and gel properties. The effect of microbial TGase on emulsions containing sodium caseinate or skim milk was studied by *Nonaka et al.*, (1992). At higher enzyme concentrations, breaking strength, strain and cohesiveness of caseinate gels decreased. *Færgemand et al.* (1997, 1999) showed that TGase induced crosslinking of sodium caseinate adsorbed at the oil-water interface increased interfacial shear viscosity by a factor of about 100 and led to a greater stability of emulsions. Storage stability of emulsions coated by crosslinked β-casein polymers also increased with increasing degree of polymerization (*Liu et al.*, 1999). Ostwald ripening of caseinate-stabilized emulsions was found to decrease when TGase was used. *Dickinson et al.*, (1999) observed that use of TGase prior to homogenization led to a modest reduction in the coarsening rate during long-term storage. In contrast, enzymatic crosslinking after homogenization increased emulsions stability for a short while, but droplet coalescence caused a complete breakdown of the emulsion during subsequent long term storage. *Kellerby et al.* (2006) was interested in determining whether TGase-crosslinked protein-stabilized oil-in-water droplet interfaces led to a reduction in lipid oxidation, but found that emulsions oxidized as rapidly as controls where interfaces had not been crosslinked. Apparently, the crosslinking did not increase resistance to mass transport or prevent access of reactive oxygen species to lipids. A variety of studies looked at the impact of TGase on the formation of emulsion-filled protein-gels (*Sharma et al.*, 2002; *Lee et al.*, 2006; *Tang et al.*, 2011; *Yang et al.*, 2011). *Tang et al.* (2011) demonstrated that soy protein isolate-stabilized emulsions formed gels with high mechanical strengths after TG treatment. Gel strength increased with increasing oil volume fraction in TGase-generated soy protein-stabilized emulsion gels (*Yang et al.*, 2011). Such enzymatically produced emulsion gels were comprised a filamentous network, that reduced loss of aroma compounds during storage (*Lee et al.*, 2006).

Another major protein fraction of bovine milk - the whey proteins – also have excellent emulsifying properties and are therefore often used in the formulation of food dispersions. Like caseins, they can be enzymatically crosslinked by TGase. Addition of TGase to α-lactalbumin, β-lactoglobulin, and blends of these two proteins at pH ranging from 6.5 to 8.0 led to the formation of large protein complexes (*Aboumahmoud et al.*, 1990; *Tanimoto et al.*, 2002). Extensive crosslinking altered their gelation properties (*Truong et al.*, 2004). It also has been reported that α-lactalbumin can be crosslinked in its native state, while β-lactoglobulin requires partial or complete denaturation to serves as a suitable substrate for an enzymatic transformation; likely due to spatial restriction in accessibility in its native state (*Eissa*, 2005).

TGase is also able to crosslink proteins derived from other sources than just milk. Examples include gluten, soya, gelatin, gum arabic, and myofibrillar proteins. TGase has also been used to generate protein-polysaccharide complexes known as coacervates (*Kurth et al.*, 1984; *Chen et al.*, 2003; *Ramírez-Suárez et al.*, 2003; *Flanagan et al.*, 2006). Typically coacervates are stabilized only by electrostatic interactions, which depend on solution composition and prevailing environmental conditions. In contrast, enzymatic formation of coacervates generates complexes that are less dependent on environmental conditions or solution composition. *Flanagan et al.* (2006) incubated mixtures of sodium caseinate and gum arabic with TGase, whereupon complexes with molecular masses of approximately 950 to 1600 kDa were formed. Gelatin-gum arabic microcapsuled were produced by complex coacervation and then crosslinked by TGase. There, the enzymatic treatment yielded results similar to those obtained when common hardening agents such as formaldehyde were used (*Dong et al.*, 2008). Finally, formation of ovomucin conjugates with casein or soy protein promoted by TGase treatment (*Kato et al.*, 1991).

**Laccase**

Laccase has gained considerable attention in the food industry as a novel agent able to stabilize particulate- or gel food structures. For example, laccase has been reported to be able to induce crosslinks in biopolymer multilayered oil-in-water emulsions (*Littoz et al.*, 2008; *Zeeb et al.*, 2012). Crosslinking occurred exclusively in the interfacial membranes and not between droplets, since no droplet-droplet aggregates were formed. During storage, size of enzymatically treated emulsions stabilized by multiple biopolymer layers decreased which was attributed to enzymatic hydrolysis (*Zeeb et al.*, 2012). When treated with laccase, sugar beet pectin used as a multilayering agent in oil-in-water emulsions Laccase-treated, improved emulsion stability, which was attributed to the development of thick layer at the oil interface (*Jung et al.*, 2012). Crosslinked pectin-coated oil droplets had similar or significantly better stability than conventional oil droplets when subjected to NaCl addition (0 – 500 mM), $CaCl_2$ addition (0 – 250 mM), and thermal processing (30 – 90 °C for 30 min). Creaming of enzyme-treated, secondary emulsions after two freeze-thaw cycles (-8 °C for 22 h; +25 °C for 2 h) was significantly reduced (*Zeeb et al.*, 2011). Similar findings were obtained by *Chen et al.* (2010). In their study, cross-linked pectin-coated oil bodies had similar or better stability than uncoated oil bodies when subjected to pH changes (3 to 7), NaCl addition (0 to 500 mM), and freeze-thaw cycling (-20 °C for 22 h; +40 °C for 2 h). Ostwald ripening of coated *n*-alkane-in-water emulsions rates after enzymatic crosslinking was studied. Ripening rates of single-layered, double-layered, and crosslinked *n*-alkane-in-water emulsions increased as the chain length of *n*-alkanes decreased. Emulsions containing crosslinked fish gelatin-beet pectin coated droplets had lower droplet growth rates ($3.1 \pm 0.3 \times 10^{-26}$ $m^3/s$) than fish gelatin-stabilized droplets ($7.3 \pm 0.2 \times 10^{-26}$ $m^3/s$), which was attributed to the formation of a protective network that counteracted the Ostwald ripening driven shrinkage (*Zeeb et al.*, 2012). Furthermore, vanillic acid modification significantly improved crosslinking efficiency in proteins thereby affecting their emulsifier properties (*Ma et al.*, 2011). The accessibility of the substrate to the enzyme may play a key role in being able to induce crosslinks - a fact that may be of particular importance in assembled complex systems. The rate of laccase-induced oxidation was slower for systems in which the emulsions had been homogenized together with the sugar beet pectin as opposed to when prepared separately (*Zaidel et al.*, 2013). The study's authors suggested that differences in the structural make-up of the emulsion system might have provided different accessabilities of the ferulic acid groups on the pectin backbone for crosslinking.

Biopolymer particles may be formed and stabilized by the addition of laccase (*Matalanis et al.*, 2011). In a study by *Chen et al.* (2012), crosslinking of WPI-beet pectin complex coacervates led to formation of rigid and elastic gels with highly organized microstructures. *Selinheimo et al.* (2008) found that laccase from *Trametes hirsuta* was able to catalyze bond formation between α-caseins and various carbohydrates. There, crosslinking was significantly enhanced by the addition of ferulic acid and *p*-coumaric acid. Similar findings were obtained by *Steffensen et al.* (2008). In their study, laccase promoted crosslinking of $\beta$-lactoglobulin and $\alpha$-casein in the presence of ferulic acid. One should not though that these reactions are quite complex because the enzyme oxidizes ferulic acid and peptide substrates simultaneously (*Mattinen et al.*, 2005; *Mattinen et al.*, 2006; *Mattinen et al.*, 2008).

Moreover, the influence of laccase on the formation and stability of hydrogels was investigated (*Figueroa-Espinoza et al.*, 1999; *Carvajal-Millan et al.*, 2005). It was shown that laccase crosslinked arabinoxylans, but that in the presence of cysteine and reduced glutathione gel formation was inhibited. In contrast, tyrosine, lysine, and oxidized glutathione had no effect on the gelation of arabinoxylans (*Figueroa-Espinoza et al.*, 1998). *Figueroa-Espinoza et al.* (1999) also investigated the influence of cysteinyl caffeic acid and caffeic acid on the gelation of arabinoxylan. The authors concluded that cysteinyl caffeic acid was oxidized while the semiquinones of ferulic acid produced by laccase were reduced. As a consequence, the gel formation of feruloylated arabinoxylans chains was inhibited since ferulic acid could not be oxidized into dimmers until all cysteinyl caffeic acids were consumed. *Kuuva et al.* (2003) and *Norsker et al.* (2000) investigated the rheological properties of laccase-induced sugar beet gels. They found that sugar beet pectin did not readily form gels in the absence of laccase, and laccase had a significant influence on the gelling rate, which also depended on environmental conditions. The degree of crosslinking could be modulated by the amount of pectin and enzyme added, and the reaction time (*Micard et al.*, 1999; *Minussi et al.*, 2002).

Electrospinning has been successfully used to immobilize laccase (from *Trametes versicolor*) in order to retain enzyme activity. *Dai et al.* (2010) showed that after 10 successive runs in the enzyme reactor, the immobilized laccase maintained 50% of its initial activity.

**Tyrosinase**

Tyrosinase catalyzes the oxidation of phenolic compounds. Subsequently produced quinones are reactive und induce crosslinking in various biopolymers (*Halaouli et al.*, 2006). The potential use of tyrosinase in food systems and food dispersions has only recently begun to be fully investigated. In these studies, a variety of food substrates such as polysaccharides and proteins were used. *Aberg et al.* (2004) for example analyzed the tyrosinase-grafting of casein hydrolysate onto the amine-containing polysaccharide chitosan. They found that the viscosity of the peptide/chitosan solution increased greatly upon use of tyrosinase. Moreover, tyrosinase has shown to be able to crosslink gelatin and chitosan leading to the formation of gelatin–chitosan co-gels (*Chen et al.*, 2001; *Chen et al.*, 2002; *Chen et al.*, 2003). When the ability of *Agaricus bisporus* tyrosinase to catalyze the oxidation of tyrosine residues of silk sericin was studied, oxidized sericin samples were able to undergo non-enzymatic coupling with chitosan (*Anghileri et al.*, 2007). Small phenolic weight molecules such as caffeic acid or chlorogenic acid can be linked to protein fibers or chitosan (*Kumar et al.*, 2000; *Jus et al.*, 2008). Tyrosinase also induced rapid gelation of polyethyleneglycol and L-DOPA to form hydrogels (*Lee et al.*, 2002).

Tyrosinase affected both texture and water-holding capacity of heated chicken breast meat homogenates, two properties that are of interest to meat product manufacturers. In the same, an improved gel formation of myofibrillar suspension in presence of NaCl and tyrosinase was reported (*Lantto*, 2007). In the presence of low molecular weight phenolics, tyrosinase may also induced polymerization of proteins such as casein, α-lactalbumin and β-lactoglobulin (*Hurrell et al.*, 1982, *Thalmann et al.*, 2002). An *Agaricus bisporus* tyrosinase induced polymerization of α-lactalbumin, and casein has been successfully crosslinked by *Pycnoporus sanguineus* tyrosinase (*Thalmann et al.*, 2002; *Halaouli et al.*, 2005). *Kato et al.* (1986) observed that the ortho-quinone tyrosinase-oxidation products were able to covalently bind to proteins via sulfhydryl linkages. Finally, tyrosinase was used in wheat doughs to modify their textural properties. The formation of 5-S-cysteinyl-DOPA and DOPA as potential crosslinkers for gluten was investigated. The amount of 5-S-cysteinyl-DOPA increased when a mushroom tyrosinase was added (*Takasaki et al.*, 1997; *Takasaki et al.*, 2001).

**Glucose oxidase**

Studies to use glucose oxidase as a modifier for food dispersions are still in their early stages. In wheat dough $H_2O_2$ may be generated - a fact that has been attributed to the action of glucose oxidase. The enzyme apparently oxidizes cysteine residues of wheat proteins to form disulphide bonds thereby altering the functionality of the proteins (*Lantto*, 2007). Glucose oxidase was also found to have a significant influence on the network formation of proteins in wheat dough. It strengthened the gluten network by promoting formation of permanent crosslinks (*Dunnewind et al.*, 2002). *Rasiah et al.* (2005) observed such a bond formation between the water-soluble albumin and globulin fraction of wheat proteins. A strengthening of wheat dough and an improvement of bread quality was observed when glucose oxidase had been added, albeit detrimental effects were reported when excessive enzyme levels were used (*Bonet et al.*, 2006).

## CONCLUSIONS AND OUTLOOK

This review has shown that crosslinking enzymes such as for example transglutaminase, laccase, tyrosinase and others may be very useful tools to modify the functionality of a wide variety of food dispersions commonly used as carriers for functional ingredients. Biopolymers used as stabilizers or emulsifiers in these systems may be modified such that their pH-, salt-, and temperature-stability increases. This is therefore a very promising research area that should be further studied, especially if new enzymes were to be discovered. However, more industrial use studies are needed. Questions of scale up, and tolerances to fluctuations in processes parameters are needed to allow food manufacturers to widely employ the technology.

Additional innovations could come from the use of combinations of enzymes or sequence of enzymes, an area which has not yet been explored. To truly modulate food structures and build new and more stable food dispersions, food scientists need to better understand the structural aspects of enzymatic actions. An important question is whether food scientists can control through the creation of specific substrate structures where the enzymatic action takes place. One could for example envision to only intra-crosslink an interfacial membrane to stabilize a base dispersion, and then to later inter-crosslink these to form networks. If two or more enzymes were used, investigations would also need to focus on interactions and environmental dependences (pH-, temperature-, and salt-stability) of the double enzyme system. There is also a need to consider potential side activities of enzymes. In many cases, authors only study the desired reaction that is the focus of the investigation. However, as shown by a few authors, enzymes sometimes catalyse also other reactions leading to some unexpected outcome such as the previously described shrinkage of a multilayered emulsion.

When talking about using combinations of enzymes, the use of hydrolysing (i.e. structure degrading) enzymes should also not be neglected. One could envision using combinations of "structure-breaking" and "structure-generating" enzymes to tailor texture. Early studies have already shown that trypsin modifies milk proteins with respect to their emulsification properties. The degree of trypsin-induced hydrolysis had a significant impact on the stability of oil-in-water emulsions exposed to different destabilizing conditions. Another common used enzyme catalyzing the breakdown of chemical bonds is chymosin. As part of the renneting process, it destabilizes the protein-based colloidal system in milk leading to an enzyme-induced curd formation.

In conclusion, the review thus demonstrates that enzymes have a significant potential to modify food structures, chief amongst them food dispersions. Once the scientific basis of structure-function relationships has been better established, food manufacturers should be able to rationally design novel food structures using enzymatic approaches.

## REFERENCES

Aberg, C. M., T. Chen, et al. (2004): Enzymatic Grafting of Peptides from Casein Hydrolysate to Chitosan. Potential for Value-Added Byproducts from Food-Processing Wastes. Journal of Agricultural and Food Chemistry 52(4): 788-793.

Aberg, C. M., T. Chen, et al. (2004): Enzymatic grafting of peptides from casein hydrolysate to chitosan. Potential for value-added pyproducts from food-processing wastes. Journal of Agricultural and Food Chemistry 52(4): 788-793.

Aboumahmoud, R. and P. Savello (1990): Crosslinking of whey protein by transglutaminase. Journal of Dairy Science 73(2): 256-263.

Aguilera, J. M. (2006): Seligman lecture 2005. Food product engineering: Building the right structures. Journal of the Science of Food and Agriculture 86(8): 1147-1155.

Ajinomoto. (2009). "ACTIVA® Transglutaminase."; www.ajinomoto.de

Anghileri, A., R. Lantto, et al. (2007): Tyrosinase-catalyzed grafting of sericin peptides onto chitosan and production of protein-polysaccharide bioconjugates. Journal of Biotechnology 127(3): 508-519.

Aoki, T., E. A. Decker, et al. (2005): Influence of environmental stresses on stability of O/W emulsions containing droplets stabilized by multilayered membranes produced by a layer-by-layer electrostatic deposition technique. Food Hydrocolloids 19(2): 209-220.

Appelqvist, I. A. M., M. Golding, et al. (2007): Emulsions as delivery systems in foods. Encapsulation and controlled release technologies in food systems, Blackwell Publishing: 41-81.

Aurbach, G. D. and W. B. Jakoby (1962): The multiple functions of thiooxidase. J. Biol. Chem. 237(2): 565-568.

Benichou, A., A. Aserin, et al. (2004): Double emulsions stabilized with hybrids of natural polymers for entrapment and slow release of active matters. Advances in Colloid and Interface Science 108-109: 29-41.

Blakistone, B. A., L. W. Aurand, et al. (1986): Association of sulhydryl oxidase and xanthine oxidase in bovine mammary tissue. Journal of Dairy Science 69: 2803 - 2809.

Bonet, A., C. M. Rosell, et al. (2006): Glucose oxidase effect on dough rheology and bread quality: A study from macroscopic to molecular level. Food Chemistry 99(2): 408-415.

Bourbonnais, R. and M. G. Paice (1990): Oxidation of non-phenolic substrates: An expanded role for laccase in lignin biodegradation. FEBS Letters 267(1): 99-102.

Branco, M. C. and J. P. Schneider (2009): Self-assembling materials for therapeutic delivery. Acta Biomaterialia 5(3): 817-831.

Bunjes, H., M. H. J. Koch, et al. (2003): Influence of emulsifiers on the crystallization of solid lipid nanoparticles. Journal of Pharmaceutical Sciences 92(7): 1509-1520.

Bunzel, M., C. Funk, et al. (2004): Semipreparative isolation of dehydrodiferulic and dehydrotriferulic acids as standard substances from maize bran. Journal of Separation Science 27(13): 1080-1086.

Bunzel, M., J. Ralph, et al. (2005): Structural elucidation of new ferulic acid-containing phenolic dimers and trimers isolated from maize bran. Tetrahedron Letters 46(35): 5845-5850.

Bunzel, M., J. Ralph, et al. (2001): Diferulates as structural components in soluble and insoluble cereal dietary fibre. Journal of the Science of Food and Agriculture 81(7): 653-660.

Carunchio, F., C. Crescenzi, et al. (2001): Oxidation of ferulic acid by laccase: identification of the products and inhibitory effects of some dipeptides. Talanta 55(1): 189-200.

Caruso, F., D. Trau, et al. (2000): Enzyme encapsulation in layer-by-layer engineered polymer multilayer capsules. Langmuir 16(4): 1485-1488.

Carvajal-Millan, E., B. Guigliarelli, et al. (2005): Storage stability of laccase induced arabinoxylan gels. Carbohydrate Polymers 59(2): 181-188.

Chen, B., H. Li, et al. (2012): Formation and microstructural characterization of whey protein isolate/beet pectin coacervations by laccase catalyzed cross-linking. LWT - Food Science and Technology 47(1): 31-38.

Chen, B., D. J. McClements, et al. (2010): Stabilization of soybean oil bodies by enzyme (Laccase) cross-linking of adsorbed beet pectin coatings. Journal of Agricultural and Food Chemistry 58(16): 9259-9265.

Chen, L., G. E. Remondetto, et al. (2006): Food protein-based materials as nutraceutical delivery systems. Trends in Food Science & Technology 17(5): 272-283.

Chen, M. C. M., C. L. Chyan, et al. (2004): Constitution of stable artificial oil bodies with triacylglycerol, phospholipid, and caleosin. Journal of Agricultural and Food Chemistry 52(12): 3982-3987.

Chen, T., H. D. Embree, et al. (2003): Enzyme-catalyzed gel formation of gelatin and chitosan: potential for in situ applications. Biomaterials 24(17): 2831-2841.

Chen, T., H. D. Embree, et al. (2002): In vitro protein-polysaccharide conjugation: Tyrosinase-catalyzed conjugation of gelatin and chitosan. Biopolymers 64(6): 292-302.

Chen, T., R. Vazquez-Duhalt, et al. (2001): Combinatorial screening for enzyme-mediated coupling. Tyrosinase-catalyzed coupling to create protein-chitosan conjugates. Biomacromolecules 2(2): 456-462.

Claus, H. (2004): Laccases: Structure, reactions, distribution. Micron 35(1-2): 93-96.

Claus, H. and H. Decker (2006): Bacterial tyrosinases. Systematic and Applied Microbiology 29(1): 3-14.

Dai, Y., J. Niu, et al. (2010): In situ encapsulation of laccase in microfibers by emulsion electrospinning: Preparation, characterization, and application. Bioresource Technology 101(23): 8942-8947.

Dalgleish, D. G. (1997): Adsorption of protein and the stability of emulsions. Trends in Food Science & Technology 8(1): 1-6.

Davis, S. S. and I. M. Walker (1987): Multiple emulsions as targetable delivery systems. Methods in Enzymology 149: 51-64.

De Jong, G. A. H. and S. J. Koppelman (2002): Transglutaminase catalyzed reactions: Impact on food applications. Journal of Food Science 67(8): 2798-2806.

De Kruif, C. G., F. Weinbreck, et al. (2004): Complex coacervation of proteins and anionic polysaccharides. Current Opinion in Colloid and Interface Science 9(5): 340-349.

De La Motte, R. S. and F. W. Wagner (1987): Aspergillus niger sulfhydryl oxidase. Biochemistry 7363-7371.

Decher, G. (1997): Fuzzy nanoassemblies: Toward layered polymeric multicomposites. Science 277(5330): 1232-1237.

Decher, G., J. D. Hong, et al. (1992): Buildup of ultrathin multilayer films by a self-assembly process: III. Consecutively alternating adsorption of anionic and cationic polyelectrolytes on charged surfaces. Thin Solid Films 210-211(PART 2): 831-835.

Desai, K. G. H. and H. J. Park (2005): Recent developments in microencapsulation of food ingredients. Drying Technology 23(7): 1361-1394.

Dickinson, E. (1997): Enzymatic crosslinking as a tool for food colloid rheology control and interfacial stabilization. Trends in Food Science and Technology 8(10): 334-339.

Dickinson, E. (1997): Enzymic crosslinking as a tool for food colloid rheology control and interfacial stabilization. Trends in Food Science and Technology 8(10): 334-339.

Dickinson, E. (1997): Properties of emulsions stabilized with milk proteins: Overview of some recent developments. Journal of Dairy Science 80(10): 2607-2619.

Dickinson, E. (2008): Interfacial structure and stability of food emulsions as affected by protein-polysaccharide interactions. Soft Matter 4(5): 932-942.

Dickinson, E., C. Ritzoulis, et al. (1999): Ostwald ripening of protein-stabilized emulsions: Effect of transglutaminase crosslinking. Colloids and Surfaces B: Biointerfaces 12(3-6): 139-146.

Dickinson, E. and Y. Yamamoto (1996): Rheology of milk protein gels and protein-stabilized emulsion gels cross-linked with transglutaminase. Journal of Agricultural and Food Chemistry 44(6): 1371-1377.

Dinsmore, A. D., M. F. Hsu, et al. (2002): Colloidosomes: Selectively permeable capsules composed of colloidal particles. Science 298(5595): 1006-1009.

Dong, Z. J., S. Q. Xia, et al. (2008): Optimization of cross-linking parameters during production of transglutaminase-hardened spherical multinuclear microcapsules by complex coacervation. Colloids and Surfaces B: Biointerfaces 63(1): 41-47.

Dunnewind, B., T. van Vliet, et al. (2002): Effect of oxidative enzymes on bulk rheological properties of wheat flour doughs. Journal of Cereal Science 36(3): 357-366.

Eissa, A. S. (2005): Enzymatic modification of whey protein gels at low pH. Raleigh, North Carolina State University.

Estroff, L. A. (2008): Introduction: Biomineralization. Chemical Reviews 108(11): 4329-4331.

Estroff, L. A. and A. D. Hamilton (2001): At the interface of organic and inorganic chemistry: Bioinspired synthesis of composite materials. Chemistry of Materials 13(10): 3227-3235.

Færgemand, M., B. S. Murray, et al. (1997): Cross-Linking of milk proteins with transglutaminase at the oil-water interface. Journal of Agricultural and Food Chemistry 45(7): 2514-2519.

Færgemand, M., B. S. Murray, et al. (1999): Cross-linking of adsorbed casein films with transglutaminase. International Dairy Journal 9(3-6): 343-346.

Figueroa-Espinoza, M.-C., M.-H. Morel, et al. (1998): Effect of lysine, tyrosine, cysteine, and glutathione on the oxidative cross-linking of feruloylated arabinoxylans by a fungal laccase. Journal of Agricultural and Food Chemistry 46(7): 2583-2589.

Figueroa-Espinoza, M. C., M. H. Morel, et al. (1999): Attempt to cross-link feruloylated arabinoxylans and proteins with a fungal laccase. Food Hydrocolloids 13(1): 65-71.

Figueroa-Espinoza, M. C. and X. Rouau (1999): Effect of cysteinyl caffeic acid, caffeic acid, and *l*-dopa on the oxidative cross-linking of feruloylated arabinoxylans by a fungal laccase. Journal of Agricultural and Food Chemistry 47(2): 497-503.

Flanagan, J., Y. Gunning, et al. (2003): Effect of cross-linking with transglutaminase on the heat stability and some functional characteristics of sodium caseinate. Food Research International 36(3): 267-274.

Flanagan, J. and H. Singh (2006): Conjugation of sodium caseinate and gum arabic catalyzed by transglutaminase. Journal of Agricultural and Food Chemistry 54(19): 7305-7310.

Flanagan, J. and H. Singh (2006): Microemulsions: A potential delivery system for bioactives in food. Critical Reviews in Food Science and Nutrition 46(3): 221-237.

Folk, J. E., J. S. Finlayson, et al. (1977): The ε-(γ)-glutamyl)lysine crosslink and the catalytic role of transglutaminases. Advances in Protein Chemistry, Academic Press. Volume 31: 1-133.

Folk, J. E., S. Il Chung, et al. (1985): [46] Transglutaminases. Methods in Enzymology, Academic Press. Volume 113: 358-375.

Garti, N. (1997): Progress in stabilization and transport phenomena of double emulsions in food applications. LWT - Food Science and Technology 30(3): 222-235.

Garti, N. (1998): A new approach to improved stability and controlled release in double emulsions, by the use of graft-comb polymeric amphiphiles. Acta Polymerica 49(10-11): 606-616.

Garti, N. and C. Bisperink (1998): Double emulsions: Progress and applications. Current Opinion in Colloid and Interface Science 3(6): 657-667.

Garti, N., A. Spernath, et al. (2005): Nano-sized self-assemblies of nonionic surfactants as solubilization reservoirs and microreactors for food systems. Soft Matter 1(3): 206-218.

Gerez, C. L., G. F. de Valdez, et al. (2012): Whey protein coating bead improves the survival of the probiotic Lactobacillus rhamnosus CRL 1505 to low pH. Letters in Applied Microbiology 54(6): 552-556

Gerrard, J. A. (2002): Protein-protein crosslinking in food: Methods, consequences, applications. Trends in Food Science & Technology 13(12): 391-399.

Gianfreda, L., F. Xu, et al. (1999): Laccases: A useful group of oxidoreductive enzymes. Bioremediation Journal 3(1): 1-25.

Gomez-Hens, A. and J. M. Fernandez-Romero (2006): Analytical methods for the control of liposomal delivery systems. TrAC - Trends in Analytical Chemistry 25(2): 167-178.

Greenberg, C., P. Birckbichler, et al. (1991): Transglutaminases: multifunctional cross-linking enzymes that stabilize tissues. FASEB J. 5(15): 3071-3077.

Gu, Y. S., E. A. Decker, et al. (2004): Influence of pH and i-carrageenan concentration on physicochemical properties and stability of b-lactoglobulin-stabilized oil-in-water emulsions. Journal of Agricultural and Food Chemistry 52(11): 3626-3632.

Gu, Y. S., E. A. Decker, et al. (2007): Formation of colloidosomes by adsorption of small charged oil droplets onto the surface of large oppositely charged oil droplets. Food Hydrocolloids 21(4): 516-526.

Gübitz, G. M. and A. C. Paulo (2003): New substrates for reliable enzymes: Enzymatic modification of polymers. Current Opinion in Biotechnology 14(6): 577-582.

Guzey, D. and D. McClements (2006): Influence of environmental stresses on o/w emulsions stabilized by β-lactoglobulin–pectin and β-lactoglobulin–pectin–chitosan membranes produced by the electrostatic layer-by-layer deposition technique. Food Biophysics 1(1): 30-40.

Guzey, D. and D. J. McClements (2006): Formation, stability and properties of multilayer emulsions for application in the food industry. Advances in Colloid and Interface Science 128-130: 227-248.

Halaouli, S., M. Asther, et al. (2005): Characterization of a new tyrosinase from *Pycnoporus* species with high potential for food technological applications. Journal of Applied Microbiology 98(2): 332-343.

Halaouli, S., M. Asther, et al. (2006): Fungal tyrosinases: New prospects in molecular characteristics, bioengineering and biotechnological applications. Journal of Applied Microbiology 100(2): 219-232.

Helgason, T., T. S. Awad, et al. (2008): Influence of polymorphic transformations on gelation of tripalmitin solid lipid nanoparticle suspensions. JAOCS, Journal of the American Oil Chemists' Society 85(6): 501-511.

Henriksen, I., G. Smistad, et al. (1994): Interactions between liposomes and chitosan. International Journal of Pharmaceutics 101(3): 227-236.

Hinz, K., T. Huppertz, et al. (2007): Influence of enzymatic cross-linking on milk fat globules and emulsifying properties of milk proteins. International Dairy Journal 17(4): 289-293.

Hoober, K. L., B. Joneja, et al. (1996): A sulfhydryl oxidase from chicken egg white. J. Biol. Chem. 271(48): 30510-30516.

Hsu, M. F., M. G. Nikolaides, et al. (2005): Self-assembled shells composed of colloidal particles: Fabrication and characterization. Langmuir 21(7): 2963-2970.

Huppertz, T., M. A. Smiddy, et al. (2007): Biocompatible micro-gel particles from cross-linked casein micelles. Biomacromolecules 8(4): 1300-1305.

Hurrell, R. F., P. A. Finot, et al. (1982): Protein-polyphenol reactions. British Journal of Nutrition 47(02): 191-211.

Ignatova, M., N. Manolova, et al. (2007): Novel antibacterial fibers of quaternized chitosan and poly(vinyl pyrrolidone) prepared by electrospinning. European Polymer Journal 43(4): 1112-1122.

Iwanaga, D., D. Gray, et al. (2008): Stabilization of soybean oil bodies using protective pectin coatings formed by electrostatic deposition. Journal of Agricultural and Food Chemistry 56(6): 2240-2245.

Iwanaga, D., D. A. Gray, et al. (2007): Extraction and characterization of oil bodies from soy beans: A natural source of pre-emulsified soybean oil. Journal of Agricultural and Food Chemistry 55(21): 8711-8716.

Janolino, V. G. and H. E. Swaisgood (1975): Isolation and characterization of sulfhydryl oxidase from bovine milk. J. Biol. Chem. 250(7): 2532-2538.

Jenning, V., K. Mäder, et al. (2000): Solid lipid nanoparticles (SLN™) based on binary mixtures of liquid and solid lipids: a 1H-NMR study. International Journal of Pharmaceutics 205(1-2): 15-21.

Jenning, V., M. Schäfer-Korting, et al. (2000): Vitamin A-loaded solid lipid nanoparticles for topical use: Drug release properties. Journal of Controlled Release 66(2-3): 115-126.

Jenning, V., A. F. Thünemann, et al. (2000): Characterisation of a novel solid lipid nanoparticle carrier system based on binary mixtures of liquid and solid lipids. International Journal of Pharmaceutics 199(2): 167-177.

Jiang, S. and S. Granick (2008): Controlling the geometry (Janus balance) of amphiphilic colloidal particles. Langmuir 24(6): 2438-2445.

Jung, J. and L. Wicker (2012): Laccase mediated conjugation of sugar beet pectin and the effect on emulsion stability. Food Hydrocolloids 28(1): 168-173.

Jus, S., V. Kokol, et al. (2008): Tyrosinase-catalysed coupling of functional molecules onto protein fibres. Enzyme and Microbial Technology 42(7): 535-542.

Karadag, A., B. Özçelik, et al. (2013): Presence of Electrostatically Adsorbed Polysaccharides Improves Spray Drying of Liposomes. Journal of Food Science 78(2): E206-E221.

Kato, A. (2002): Industrial applications of maillard-type protein-polysaccharide conjugates. Food Science and Technology Research 8(3): 193-199.

Kato, A., T. Wada, et al. (1991): Ovomucin-food protein conjugates prepared through the transglutaminase reaction. Agricultural and biological chemistry 55(4): 1027-1031.

Kato, T., S. Ito, et al. (1986): Tyrosinase-catalyzed binding of 3,4-dihydroxyphenylalanine with proteins through the sulfhydryl group. Biochimica et Biophysica Acta (BBA) - General Subjects 881(3): 415-421.

Kellerby, S. S., S. G. Yeun, et al. (2006): Lipid oxidation in a menhaden oil-in-water emulsion stabilized by sodium caseinate cross-linked with transglutaminase. Journal of Agricultural and Food Chemistry 54(26): 10222-10227.

Kennedy, J. F. and C. J. Knill (1996): Food enzymes: structure and mechanism: Carbohydrate Polymers 29(4): 369-369.

Kim, J. W., A. Fernández-Nieves, et al. (2007): Colloidal assembly route for responsive colloidosomes with tunable permeability. Nano Letters 7(9): 2876-2880.

Kona, R. P., N. Qureshi, et al. (2001): Production of glucose oxidase using *Aspergillus* niger and corn steep liquor. Bioresource Technology 78(2): 123-126.

Koynova, R. and M. Caffrey (1998): Phases and phase transitions of the phosphatidylcholines. Biochimica et Biophysica Acta - Reviews on Biomembranes 1376(1): 91-145.

Kriegel, C., A. Arrechi, et al. (2008): Fabrication, functionalization, and application of electrospun biopolymer nanofibers. Critical Reviews in Food Science and Nutrition 48(8): 775-797.

Kumar, G., J. F. Bristow, et al. (2000): Enzymatic gelation of the natural polymer chitosan. Polymer 41(6): 2157-2168.

Kurth, L. and P. J. Rogers (1984): Transglutaminase catalyzed cross-linking of myosin to soya protein, casein and gluten. Journal of Food Science 49(2): 573-576.

Kusakabe, H., A. Kuninaka, et al. (1982): Purification and properties of a new enzyme, glutathione oxidase from Penicillium sp. K-6-5. Agriculture Biol. Chem. 46: 2057–2067.

Kuuva, T., R. Lantto, et al. (2003): Rheological properties of laccase-induced sugar beet pectin gels. Food Hydrocolloids 17(5): 679-684.

Lantto, R. (2007): Protein cross-linking with oxidative enzymes and transglutaminase: Effects in meat protein systems.

Laye, C., D. J. McClements, et al. (2008): Formation of biopolymer-coated liposomes by electrostatic deposition of chitosan. Journal of Food Science 73(5): N7-N15.

Lee, B. P., J. L. Dalsin, et al. (2002): Synthesis and Gelation of DOPA-Modified Poly(ethylene glycol) Hydrogels. Biomacromolecules 3(5): 1038-1047.

Lee, H. A., S. J. Choi, et al. (2006): Characteristics of sodium caseinate- and soy protein isolate-stabilized emulsion-gels formed by microbial transglutaminase. Journal of Food Science 71(6): C352-C357.

Littoz, F. and D. J. McClements (2008): Bio-mimetic approach to improving emulsion stability: Cross-linking adsorbed beet pectin layers using laccase. Food Hydrocolloids 22(7): 1203-1211.

Liu, M. and S. Damodaran (1999): Effect of transglutaminase-catalyzed polymerization of *b*-casein on its emulsifying properties. Journal of Agricultural and Food Chemistry 47(4): 1514-1519.

Lombardi Borgia, S., M. Regehly, et al. (2005): Lipid nanoparticles for skin penetration enhancement-correlation to drug localization within the particle matrix as determined by fluorescence and parelectric spectroscopy. Journal of Controlled Release 110(1): 151-163.

Ma, H., P. Forssell, et al. (2011): Improving laccase catalyzed cross-linking of whey protein isolate and their application as emulsifiers. Journal of Agricultural and Food Chemistry 59(4): 1406-1414.

Maherani, B., E. Arab-Tehrany, et al. (2011): Liposomes: A review of manufacturing techniques and targeting strategies. Current Nanoscience 7(3): 436-452.

Manski, J. M., A. J. van der Goot, et al. (2007): Influence of shear during enzymatic gelation of caseinate-water and caseinate-water-fat systems. Journal of Food Engineering 79(2): 706-717.

Matalanis, A., O. G. Jones, et al. (2011): Structured biopolymer-based delivery systems for encapsulation, protection, and release of lipophilic compounds. Food Hydrocolloids 25(8): 1865-1880.

Matheis, G. and J. R. Whitaker (1984): Peroxidase-catalyzed cross linking of proteins. Journal of Protein Chemistry 3(1): 35-48.

Matheis, G. and J. R. Whitaker (1987): A review: Enzymatic cross-linking of proteins applicable to foods. J. Food Biochem. 11(4): 309-327.

Mattinen, M.-L., M. Hellman, et al. (2006): Effect of protein structure on laccase-catalyzed protein oligomerization. Journal of Agricultural and Food Chemistry 54(23): 8883-8890.

Mattinen, M.-L., M. Hellman, et al. (2008): Laccase and tyrosinase catalysed polymerization of proteins and peptides. Journal of Biotechnology 136(Supplement 1): S318-S318.

Mattinen, M.-L., K. Kruus, et al. (2005): Laccase-catalyzed polymerization of tyrosine-containing peptides. FEBS Journal 272(14): 3640-3650.

Mayer, A. M. and R. C. Staples (2002): Laccase: New functions for an old enzyme. Phytochemistry 60(6): 551-565.

McClements, D. J. (2004): Food emulsions: Principles, practice, and techniques. Boca Raton, CRC Press.

McClements, D. J. (2006): Non-covalent interactions between proteins and polysaccharides. Biotechnology Advances 24(6): 621-625.

McClements, D. J. and E. A. Decker (2000): Lipid oxidation in oil-in-water emulsions: Impact of molecular environment on chemical reactions in heterogeneous food systems. Journal of Food Science 65(8): 1270-1282.

McClements, D. J., E. A. Decker, et al. (2009): Structural design principles for delivery of bioactive components in nutraceuticals and functional foods. Critical Reviews in Food Science and Nutrition 49(6): 577 - 606.

McClements, D. J., E. A. Decker, et al. (2007): Emulsion-based delivery systems for lipophilic bioactive components. Journal of Food Science 72(8): R109-R124.

McClements, D. J. and K. Demetriades (1998): An integrated approach to the development of reduced-fat food emulsions. Critical Reviews in Food Science and Nutrition 38(6): 511-536.

McClements, D. J. and J. Rao (2011): Food-grade nanoemulsions: Formulation, fabrication, properties, performance, biological fate, and potential toxicity. Critical Reviews in Food Science and Nutrition 51(4): 285-330.

McClements, D. J., K. Stefan, et al. (2009): Biopolymers in food emulsions. Modern Biopolymer Science. San Diego, Academic Press: 129-166.

Mehnert, W. and K. Mäder (2001): Solid lipid nanoparticles: Production, characterization and applications. Advanced Drug Delivery Reviews 47(2-3): 165-196.

Micard, V. and J. F. Thibault (1999): Oxidative gelation of sugar-beet pectins: Use of laccases and hydration properties of the cross-linked pectins. Carbohydrate Polymers 39(3): 265-273.

Minussi, R. C., G. M. Pastore, et al. (2002): Potential applications of laccase in the food industry. Trends in Food Science & Technology 13(6-7): 205-216.

Mota, M. and J. Empis (2000): Novel foods and food ingredients: What is the mission of scientists and technologists? Trends in Food Science & Technology 11(4-5): 161-168.

Motoki, M. and K. Seguro (1998): Transglutaminase and its use for food processing. Trends in Food Science & Technology 9(5): 204-210.

Muller, R. H. and C. M. Keck (2004): Challenges and solutions for the delivery of biotech drugs: A review of drug nanocrystal technology and lipid nanoparticles. Journal of Biotechnology 113(1-3): 151-170.

Müller, R. H., M. Radtke, et al. (2002): Nanostructured lipid matrices for improved microencapsulation of drugs. International Journal of Pharmaceutics 242(1-2): 121-128.

Müller, R. H., M. Radtke, et al. (2002): Solid lipid nanoparticles (SLN) and nanostructured lipid carriers (NLC) in cosmetic and dermatological preparations. Advanced Drug Delivery Reviews 54, Supplement(0): S131-S155.

Murphy, D. J., I. Hernández-Pinzón, et al. (2001): Role of lipid bodies and lipid-body proteins in seeds and other tissues. Journal of Plant Physiology 158(4): 471-478.

Murphy, D. J., I. Hernendez-Pinzon, et al. (2000): New insights into the mechanisms of lipid-body biogenesis in plants and other organisms. Biochemical Society Transactions 28(6): 710-711.

Muschiolik, G. (2007): Multiple emulsions for food use. Current Opinion in Colloid & Interface Science 12(4-5): 213-220.

Myllärinen, P., J. Buchert, et al. (2007): Effect of transglutaminase on rheological properties and microstructure of chemically acidified sodium caseinate gels. International Dairy Journal 17(7): 800-807.

Niku-Paavola, M. L. and L. Viikari (2000): Enzymatic oxidation of alkenes. Journal of Molecular Catalysis B: Enzymatic 10(4): 435-444.

Nio, N., M. Motoki, et al. (1986): Gelation mechanism of protein solution by transglutaminase. Agric. Biol. Chem. 50(4): 851-855.

Noble, P. F., O. J. Cayre, et al. (2004): Fabrication of "hairy" colloidosomes with shells of polymeric microrods. Journal of the American Chemical Society 126(26): 8092-8093.

Nonaka, M., H. Sakamoto, et al. (1992): Sodium caseinate and skim milk gels formed by incubation with microbial transglutaminase. Journal of Food Science 57(5): 1214-1241.

Norsker, M., M. Jensen, et al. (2000): Enzymatic gelation of sugar beet pectin in food products. Food Hydrocolloids 14(3): 237-243.

O'Connell, J. E. and C. G. de Kruif (2003): b-Casein micelles; cross-linking with transglutaminase. Colloids and Surfaces A: Physicochemical and Engineering Aspects 216(1-3): 75-81.

Octavio, L. C., P. P. Ma, et al. (2006): Laccases. Advances in Agricultural and Food Biotechnology. R. G. Guevara-González and I. Torres-Pacheco.

Ostrowski, M. C. and W. S. Kistler (2002): Properties of a flavoprotein sulfhydryl oxidase from rat seminal vesicle secretion. Biochemistry 19(12): 2639-2645.

Palanuwech, J. and J. N. Coupland (2003): Effect of surfactant type on the stability of oil-in-water emulsions to dispersed phase crystallization. Colloids and Surfaces A: Physicochemical and Engineering Aspects 223(1 - 3): 251-262.

Partanen, R., A. Paananen, et al. (2009): Effect of transglutaminase-induced cross-linking of sodium caseinate on the properties of equilibrated interfaces and foams. Colloids and Surfaces A: Physicochemical and Engineering Aspects 344(1-3): 79-85.

Peng, C. C., I. P. Lin, et al. (2003): Size and stability of reconstituted sesame oil bodies. Biotechnology Progress 19(5): 1623-1626.

Ramírez-Suárez, J. C. and Y. L. Xiong (2003): Effect of transglutaminase-induced cross-linking on gelation of myofibrillar/soy protein mixtures. Meat Science 65(2): 899-907.

Rasiah, I. A., K. H. Sutton, et al. (2005): Crosslinking of wheat dough proteins by glucose oxidase and the resulting effects on bread and croissants. Food Chemistry 89(3): 325-332.

Rescigno, A., F. Sollai, et al. (2002): Tyrosinase inhibition: General and applied aspects. Journal of Enzyme Inhibition and Medicinal Chemistry 17(4): 207-218.

Rodríguez-Abreu, C. and M. Lazzari (2008): Emulsions with structured continuous phases. Current Opinion in Colloid and Interface Science 13(4): 198-205.

Rodríguez Couto, S. and J. L. Toca Herrera (2006): Industrial and biotechnological applications of laccases: A review. Biotechnology Advances 24(5): 500-513.

Rossier-Miranda, F. J., C. G. P. H. Schroën, et al. (2009): Colloidosomes: Versatile microcapsules in perspective. Colloids and Surfaces A: Physicochemical and Engineering Aspects 343(1-3): 43-49.

Salminen, H. and J. Weiss (2013): Electrostatic adsorption and stability of whey protein–pectin complexes on emulsion interfaces. Food Hydrocolloids(0): 1 - 10; Article in Press.

Sánchez-Ferrer, Á., J. Neptuno Rodríguez-López, et al. (1995): Tyrosinase: a comprehensive review of its mechanism. Biochimica et Biophysica Acta (BBA) - Protein Structure and Molecular Enzymology 1247(1): 1-11.

Sanguansri, P. and M. A. Augustin (2006): Nanoscale materials development - a food industry perspective. Trends in Food Science & Technology 17(10): 547-556.

Saraf, S., R. Rathi, et al. Colloidosomes an advanced vesicular system in drug delivery. Asian Journal of Scientific Research 4(1): 1-15.

Schmitt, C., C. Sanchez, et al. (1998): Structure and technofunctional properties of protein-polysaccharide complexes: A review. Critical Reviews in Food Science and Nutrition 38(8): 689-753.

Schorsch, C., H. Carrie, et al. (2000): Cross-linking casein micelles by a microbial transglutaminase: influence of cross-links in acid-induced gelation. International Dairy Journal 10(8): 529-539.

Schubert, M. A. and C. C. Müller-Goymann (2005): Characterisation of surface-modified solid lipid nanoparticles (SLN): Influence of lecithin and nonionic emulsifier. European Journal of Pharmaceutics and Biopharmaceutics 61(1-2): 77-86.

Selinheimo, E. (2008): Tyrosinase and Laccase as Novel Crosslinking Tools for Food Biopolymers. Finland, Helsinki University of Technology.

Selinheimo, E., P. Lampila, et al. (2008): Formation of protein-oligosaccharide conjugates by laccase and tyrosinase. Journal of Agricultural and Food Chemistry 56(9): 3118-3128.

Selinheimo, E., M. Saloheimo, et al. (2006): Production and characterization of a secreted, C-terminally processed tyrosinase from the filamentous fungus <i>Trichoderma reesei</i>. FEBS Journal 273(18): 4322-4335.

Sharma, R., M. Zakora, et al. (2002): Characteristics of oil-water emulsions stabilized by an industrial a-lactalbumin concentrate, cross-linked before and after emulsification, by a microbial transglutaminase. Food Chemistry 79(4): 493-500.

Shimada, T. L. and I. Hara-Nishimura (2010): Oil-body-membrane proteins and their physiological functions in plants. Biological and Pharmaceutical Bulletin 33(3): 360-363.

Siekmann, B. and K. Westesen (1994): Thermoanalysis of the recrystallization process of melt-homogenized glyceride nanoparticles. Colloids and Surfaces B: Biointerfaces 3(3): 159-175.

Simovic, S. and C. A. Prestidge (2008): Colloidosomes from the controlled interaction of submicrometer triglyceride droplets and hydrophilic silica nanoparticles. Langmuir 24(14): 7132-7137.

Sliwkowski, M. X., H. E. Swaisgood, et al. (1984): Kinetic mechanism and specificity of bovine milk sulphydryl oxidase. Biochem J 1: 51 - 55.

Smiddy, M. A., J. E. G. H. Martin, et al. (2006): Stability of casein micelles cross-linked by transglutaminase. Journal of Dairy Science 89(6): 1906-1914.

Smith, A. T. and N. C. Veitch (1998): Substrate binding and catalysis in heme peroxidases. Current Opinion in Chemical Biology 2(2): 269-278.

Souto, E. B. and R. H. Müller (2005): SLN and NLC for topical delivery of ketoconazole. Journal of Microencapsulation 22(5): 501-510.

Souto, E. B. and R. H. Müller (2006): Investigation of the factors influencing the incorporation of clotrimazole in SLN and NLC prepared by hot high-pressure homogenization. Journal of Microencapsulation 23(4): 377-388.

Steffensen, C. L., M. L. Andersen, et al. (2008): Cross-linking proteins by laccase-catalyzed oxidation: Importance relative to other modifications. Journal of Agricultural and Food Chemistry 56(24): 12002-12010.

Sukhorukov, G. B., E. Donath, et al. (1998): Stepwise polyelectrolyte assembly on particle surfaces: A novel approach to colloid design. Polymers for Advanced Technologies 9(10-11): 759-767.

Surh, J., Y. S. Gu, et al. (2005): Influence of environmental stresses on stability of O/W emulsions containing cationic droplets stabilized by SDS-fish gelatin membranes. Journal of Agricultural and Food Chemistry 53(10): 4236-4244.

Takasaki, S. and S. Kawakishi (1997): Formation of protein-bound 3,4-dihydroxyphenylalanine and 5-S-cysteinyl-3,4-dihydroxyphenylalanine as new cross-linkers in gluten. Journal of Agricultural and Food Chemistry 45(9): 3472-3475.

Takasaki, S., S. Kawakishi, et al. (2001): Polymerisation of gliadin mediated by mushroom tyrosinase. Lebensmittel-Wissenschaft und-Technologie 34(8): 507-512.

Tang, C. H., L. Chen, et al. (2011): Mechanical and water-holding properties and microstructures of soy protein isolate emulsion gels induced by $CaCl_2$, glucono-d-lactone (GDL), and transglutaminase: Influence of thermal treatments before and/or after emulsification. Journal of Agricultural and Food Chemistry 59(8): 4071-4077.

Tanimoto, S. Y. and J. E. Kinsella (2002): Enzymic modification of proteins: Effects of transglutaminase cross-linking on some physical properties of b-lactoglobulin. Journal of Agricultural and Food Chemistry 36(2): 281-285.

Taylor, T. M., P. M. Davidson, et al. (2005): Liposomal nanocapsules in food science and agriculture. Critical Reviews in Food Science and Nutrition 45(7-8): 587-605.

Thalmann, C. R. and T. Lötzbeyer (2002): Enzymatic cross-linking of proteins with tyrosinase. European Food Research and Technology 214(4): 276-281.

Thorpe, C., K. L. Hoober, et al. (2002): Sulfhydryl oxidases: emerging catalysts of protein disulfide bond formation in eukaryotes. Archives of Biochemistry and Biophysics 405(1): 1-12.

Thongkaew, C., and Weiss, J. (2011): Modulation of segregative biopolymers interaction: The effect of sequential changes in pH and ionic strength. Delivery of Functionality in Complex Food Systems; Physically-Inspired Approaches from the Nanoscale to the Microscale, 4th International Symposium, 2011. Guelph, Canada.

Thongkaew, C., and Weiss, J. (2011): Modulation of segregative biopolymers interaction by pH and ionic strength. Delivery of Functionality in Complex Food Systems; Physically-Inspired Approaches from the Nanoscale to the Microscale, 4th International Symposium, 2011. Guelph, Canada.

Thongkaew, C., Gibis, M., and Weiss, J. (2013): Complex coacervation stabilizes proteins against precipitation by polyphenols; IFT Annual Meeting and Food Expo, 2013. Chicago, USA.

Thurston, C. F. (1994): The structure and function of fungal laccases. Microbiology 140(1): 19-26.

Truong, V. D., D. A. Clare, et al. (2004): Cross-Linking and rheological changes of whey proteins treated with microbial transglutaminase. Journal of Agricultural and Food Chemistry 52(5): 1170-1176.

Tzen, J. T. C., Y. Z. Cao, et al. (1993): Lipids, proteins, and structure of seed oil bodies from diverse species. Plant Physiology 101(1): 267-276.

Tzen, J. T. C. and A. H. C. Huang (1992): Surface structure and properties of plant seed oil bodies. Journal of Cell Biology 117(2): 327-335.

van Dijken, J. P. and M. Veenhuis (1980): Cytochemical localization of glucose oxidase in peroxisomes of Aspergillus niger. Applied Microbiology and Biotechnology 9: 275-283.

Veitch, N. C. (2004): Horseradish peroxidase: A modern view of a classic enzyme. Phytochemistry 65(3): 249-259.

Videira, M. A., M. F. Botelho, et al. (2002): Lymphatic uptake of pulmonary delivered radiolabelled solid lipid nanoparticles. Journal of Drug Targeting 10(8): 607-613.

Wang, Y. and W. Wu (2006): In situ evading of phagocytic uptake of stealth solid lipid nanoparticles by mouse peritoneal macrophages. Drug Delivery 13(3): 189-192.

Weiss, J., E. A. M. Decker, J., et al. (2008): Solid lipid nanoparticles as delivery systems for bioactive food components. Food Biophysics 3: 146-154.

Weiss, J., S. Gaysinsky, et al. (2009): Nanostructured encapsulation systems: Food antimicrobials. Global Issues in Food Science and Technology, Academic Press, New York.

Weiss, J., P. Takhistov, et al. (2006): Functional materials in food nanotechnology. Journal of Food Science 71(9).

Were, L. M., B. D. Bruce, et al. (2003): Size, stability, and entrapment efficiency of phospholipid nanocapsules containing polypeptide antimicrobials. Journal of Agricultural and Food Chemistry 51(27): 8073-8079.

Westesen, K., H. Bunjes, et al. (1997): Physicochemical characterization of lipid nanoparticles and evaluation of their drug loading capacity and sustained release potential. Journal of Controlled Release 48(2-3): 223-236.

Westesen, K., B. Siekmann, et al. (1993): Investigations on the physical state of lipid nanoparticles by synchrotron radiation X-ray diffraction. International Journal of Pharmaceutics 93(1-3): 189-199.

Whitaker, J. R. and G. Matheis (1984): Modification of proteins by polyphenol oxidase and peroxidase and their products. Journal of Food Biochemistry 8(3): 137-162.

Wilkinson, B. and H. F. Gilbert (2004): Protein disulfide isomerase. Biochimica et Biophysica Acta (BBA) - Proteins & Proteomics 1699(1-2): 35-44.

Wissing, S. A., O. Kayser, et al. (2004): Solid lipid nanoparticles for parenteral drug delivery. Advanced Drug Delivery Reviews 56(9): 1257-1272.

Wörle, G., B. Siekmann, et al. (2006): Transformation of vesicular into cubic nanoparticles by autoclaving of aqueous monoolein/poloxamer dispersions. European Journal of Pharmaceutical Sciences 27(1): 44-53.

Yang, M., F. Liu, et al. (2011): Properties and microstructure of transglutaminase-set soy protein-stabilized emulsion gels. Food Research International.

Yaqoob Khan, A., S. Talegaonkar, et al. (2006): Multiple Emulsions: An Overview. Current Drug Delivery 3: 429-443.

Yaropolov, A. I., O. V. Skorobogat'ko, et al. (1994): Laccase - Properties, catalytic mechanism, and applicability. Applied Biochemistry and Biotechnology 49(3): 257-280.

Yazan, Y., M. Seiller, et al. (1993): Multiple emulsions. Boll Chim Farma 132(6): 187-196.

Yuan, Z. Y. and T. J. Jiang (2003): Horseradish peoxidase. Handbook of food enzymology. J.R. Whitaker, A.G.J. Voragen and D. W. S. Wong. Basel, Marcel Dekker Inc: 408.

Zaidel, D. N. A., I. S. Chronakis, et al. (2013): Stabilization of oil-in-water emulsions by enzyme catalyzed oxidative gelation of sugar beet pectin. Food Hydrocolloids 30(1): 19-25.

Zeeb, B., L. Fischer, et al. (2011): Cross-linking of interfacial layers affects the salt and temperature stability of multilayered emulsions consisting of fish gelatin and sugar beet pectin. Journal of Agricultural and Food Chemistry 59(19): 10546-10555.

Zeeb, B., M. Gibis, et al. (2012): Crosslinking of interfacial layers in multilayered oil-in-water emulsions using laccase: Characterization and pH-stability. Food Hydrocolloids 27(1): 126-136.

Zeeb, B., M. Gibis, et al. (2012): Influence of interfacial properties on Ostwald ripening in crosslinked multilayered oil-in-water emulsions. Journal of Colloid and Interface Science 387(1): 65-73.

Zhang, L., S. Hou, et al. (2004): Uptake of folate-conjugated albumin nanoparticles to the SKOV3 cells. International Journal of Pharmaceutics 287(1-2): 155-162.

Zhang, Y., T. L. Chwee, et al. (2005): Recent development of polymer nanofibers for biomedical and biotechnological applications. Journal of Materials Science: Materials in Medicine 16(10): 933-946.

Zhu, Y., A. Rinzema, et al. (1995): Microbial transglutaminase - A review of its production and application in food processing. Applied Microbiology and Biotechnology 44(3-4): 277-282.

# CHAPTER 2

## Crosslinking of interfacial layers in multilayered oil-in-water emulsions using laccase: Characterization and pH-stability

*Benjamin Zeeb[1], Monika Gibis[1], Lutz Fischer[2], Jochen Weiss[1]*

[1] Department of Food Physics and Meat Science, University of Hohenheim, Garbenstrasse 21/25, 70599 Stuttgart, Germany

[2] Department of Food Biotechnology, University of Hohenheim, Garbenstrasse 25,70599 Stuttgart, Germany

Reprinted from "*Crosslinking of interfacial layers in multilayered oil-in-water emulsions using laccase: Characterization and pH-stability*", Zeeb, B., Gibis, M., Fischer, L., Weiss, J., Food Hydrocolloids, 2012, 27(1), p. 126-136 with permission from Elsevier.

## ABSTRACT

The enzymatic crosslinking of polymer layers adsorbed at the interface of oil-in-water emulsions was investigated. A sequential two step process, based on the electrostatic deposition of pectin onto a fish gelatin interfacial membrane was used to prepare emulsions containing oil droplets stabilized by fish gelatin – beet pectin – membranes (citrate buffer, 10 mM, pH 3.5). First, a fine dispersed primary emulsion (5% soybean oil (w/v), 1% (w/w) gelatin solution (citrate buffer, 10 mM, pH 3.5) was produced using a high pressure homogenizer. Second, a series of secondary emulsions were formed by diluting the primary emulsion into pectin solutions (0 – 0.4% (w/w)) to coat the droplets. Oil droplets of stable emulsions with different oil droplet concentrations (0.1%, 0.5%, 1.0% (w/v)) were subjected to enzymatic crosslinking. Laccase was added to the fish gelatin – beet pectin emulsions and emulsions were incubated for 15 min at room temperature. The pH- and storage stability of primary, secondary and secondary, laccase-treated emulsions was determined. Results indicated that crosslinking occurred exclusively in the layers and not between droplets, since no aggregates were formed. Droplet size increased from 350 to 400 nm regardless of oil droplet concentrations within a matter of minutes after addition of laccase suggesting formation of covalent bonds between pectin adsorbed at interfaces and pectin in the aqueous phase in the vicinity of droplets. During storage, size of enzymatically treated emulsions decreased, which was found to be due to enzymatic hydrolysis. Results suggest that biopolymer-crosslinking enzymes could be used to enhance stability of multilayered emulsions.


**Keywords:** Multilayered emulsions; Gelatin; Sugar beet pectin; Laccase; Stability

## INTRODUCTION

Traditionally, oil-in-water (O/W) emulsions are produced by homogenizing lipids with an aqueous emulsifier solution (*Guzey et al.*, 2006). Emulsifiers adsorb to freshly formed surfaces, reduce surface tension which promotes further disruption of oil droplets, and forms a membrane surrounding the droplets to prevent from aggregating and/or flocculating (*McClements*, 2004). Proteins are widely used to stabilize O/W emulsions because of their ability to provide both high steric and electrostatic repulsive forces (*Tornberg et al.*, 1977; *McClements*, 2004; *Dalgleish*, 2006). However, the stability of protein-stabilized emulsions depends on pH and ionic strength, and protein-stabilized emulsions often tend to destabilize during dehydration, freezing, and/or chilling (*McClements*, 2004; *Guzey et al.*, 2006).

Physical, chemical or enzymatic crosslinking of biopolymers in food emulsions has been reported to enhance their stability and functionality. For example, use of naturally-occurring reaction that led to chemical crosslinking enhanced the functional properties of protein-stabilized emulsions (*Romoscanu et al.*, 2005). There, proteins at oil-water interfaces were first heated to promote crosslinking between the individual molecules. A subsequent spray drying step without addition of any hydrocolloids led to the formation of a stable powder composed of oil droplets embedded in a crosslinked protein network (*Romoscanu et al.*, 2006; *Mezzenga et al.*, 2010). Moreover, coating of O/W emulsions by layer-by-layer (LbL) electrostatic deposition technique with charged biopolymers has shown to improve resistance of emulsions to environmental stresses (*Gu et al.*, 2004; *Ogawa et al.*, 2004; *Gu et al.*, 2005). Such called "multilayered" emulsions can be formed by alternatingly adding oppositely (to the droplet interface) charged proteins or polysaccharides at appropriate concentrations (*Aoki et al.*, 2005; *Gu et al.*, 2005; *Guzey et al.*, 2006). Both, pH and ionic strength play a key role in the assembly of multilayered emulsions since they influence the degree of ionization of the charged functional groups on the respective biopolymers (e.g. amino and carboxy groups). The charge of the biopolymers in turn affects the electrostatic interactions responsible for the attractive forces that lead to the deposition of biopolymers onto the droplets´ surfaces (*Schmitt et al.*, 1998).

However, changes in the environmental conditions after deposition of biopolymer layers may lead to desorption of multilayers and destabilization of multilayered emulsions (*Guzey et al.*, 2006). We hypothesize that the formation of covalent bonds between biopolymers within the multiple layers of emulsions could prevent such an environmentally-induced destabilization. A wide variety of enzymes could potentially be used to stabilize deposited biopolymer membranes (*Gübitz et al.*, 2003). To date, one of the most commonly industrially used crosslinking enzyme is transglutaminase. Transglutaminase (TGase) is known to crosslink protein residues such as whey protein isolates (WPI), caseinates and gelatin via γ-carboxyl (glutamine)-ε-amino (lysine) isopeptide bond formation. *Hernàndez-Balaba et al.* (2009) demonstrated the formation of gel networks between WPI and gelatin after TGase treatment. The TGase-catalyzed crosslinking of β-casein micelles eliminated the ability of micelles to dissociate on cooling and disruption, and extensive polymerization of β-lactoglobulin resulted in the formation of a weak gel (*Tanimoto et al.*, 2002; *O'Connell et al.*, 2003; *Smiddy et al.*, 2006).

More recently, other classes of enzymes have been explored for their ability to crosslink biopolymers, noticeably laccases. Laccases belong to the group of oxidoreductases with a multinuclear copper-containing active-site (*Riva*, 2006). They have been used to catalyze the oxidation of phenol derivates in a two step radical mechanism like ferulic acid groups (*Matheis et al.*, 1987; *Uyama et al.*, 2002). Several studies have demonstrated that oxidation of tyrosine and tyrosine residues in proteins with laccase were possible (*Mattinen et al.*, 2005; *Mattinen et al.*, 2008). Fungal laccase was able to induce the formation of oligomers and polymers of whey protein isolates (WPI) in the presence of chlorogenic acid, but no gelling of WPI solutions with 10 – 20% protein was observed (*Færgemand et al.*, 1998). Chemically modified and enzymatically treated WPI was shown to enhance the storage stability of WPI-stabilized emulsions (*Ma et al.*, 2011). Laccase-induced crosslinking of sugar beet pectin in multilayered emulsions improves NaCl-stability of these emulsions (*Littoz et al.*, 2008).

The objective of the current study was to demonstrate the preparation of stable O/W emulsions stabilized by multilayer interfacial membranes made from food-grade biopolymers and to analyze if these emulsions have functional properties or improved stability to environmental stresses after enzyme treatment. Laccase was used to crosslink pectin molecules via ferulic acid bonds adsorbed to the surface of oil droplets. We hypothesized that laccase-treated emulsions may have better resistance against pH-triggered destabilization than primary and secondary emulsions.

## MATERIALS AND METHODS

**Materials.** Sugar beet pectin (#1 09 03 135), apple pectin (#0 09 11 045), and citrus pectin (#0 09 06 501) were donated by Herbstreith & Fox KG (Neuenbürg, Germany) and used without further purification. As stated by the manufacturer the degree of esterification of the beet pectin, apple pectin, and citrus pectin was 55%, 38%, and 71%, respectively. Cold water fish skin gelatin (#049K0050) was purchased from Sigma-Aldrich Co. (Steinheim, Germany). Its average molecular weight and p*I* value were reported to be ca. 60 kDa and pH 6, respectively. Laccase (#0001437590, from *Trametes versicolor*) was obtained from Sigma-Aldrich Co. (Steinheim, Germany). The Laccase obtained was reported to have 20.7 Units per mg (AU) of enzyme. Citric acid monohydrate (#409107294, purity ≥ 99.5%) was obtained from Carl Roth GmbH & Co. KG (Karlsruhe, Germany) and sodium citrate dihydrate (#26996TH, purity ≥ 99.0%) was purchased from SAFC (St. Louis, MO). Soybean oil was obtained from a local supermarket and was used without further purification. *Trans*-ferulic acid (#46278, purity ≥ 99.0%) and hydroxybenzoic acid (#54610, purity ≥ 99.0%) were obtained from Sigma-Aldrich Co. (Steinheim, Germany). Analytical grade hydrochloric acid (HCl), sodium hydroxide (NaOH), ethyl acetate and acetonitrile were purchased from Carl Roth GmbH & Co. KG (Karlsruhe, Germany). Distilled water was used for the preparation of all samples.

**Solution preparation.** Aqueous emulsifier solutions were prepared by dispersing 1% (w/w) fish gelatin powder into 10 mM citrate buffer (pH 3.5). Sugar beet pectin solutions were prepared by dispersing 2% (w/w) powdered pectin into 10 mM citrate buffer at pH 3.5 followed by stirring overnight to ensure complete hydration. pH was then adjusted to 3.5 using 1 M HCl and/or 1 M NaOH. Enzyme solutions were prepared by dispersing enzyme powder into 10 mM citrate buffer (pH 3.5) followed by stirring for 30 min. Different enzyme/beet pectin ratios were tested, namely 0.048 mg/4 mg (1 AU), 0.24 mg/4 mg (5 AU), 0.48 mg/4 mg (10 AU), and 2.4 mg/4 mg (50 AU).

**Characterization of base polymer solutions.**

*Influence of pH on ζ-potential.* Sugar beet pectin and fish gelatin solutions (1.0% (w/w), 10 mM citrate buffer) were prepared and then adjusted to pH 3 – 10 using 1 M HCl and/or 1 M NaOH. All polymer solutions were kept for 1 minute after reaching the pH value before transferring (5 ml) into glass test tubes and analyzing ζ-potential after 24 h. ζ-potential of pectin and gelatin solutions were then plotted as a function of pH.

*Laccase-catalyzed cross-linking.* UV-visible measurements of laccase-treated sugar beet pectin and fish gelatin solutions were investigated. Polymer solutions (0.1% (w/w)) were prepared in 10 mM citrate buffer at pH 3.5 and laccase (0 – 50 AU) was added. The oxidation of ferulic acid was followed by measuring the decrease in absorbance at 325 nm at 25 °C for 1 hour using an UV-visible light spectrophotometer (HP 8453, Agilent with application software Chemstation Agilent Technologies 95-00). Buffer solutions containing no polymer were used as blanks. Additionally, rheological investigations of laccase-treated polymer solutions were conducted. Solutions of beet pectin and fish gelatin (0.5% (w/w)) were prepared in 10 mM citrate buffer at pH 3.5 and laccase was added and followed by stirring for 15 min. After stirring, the flow behaviour of the gelatin and pectin solutions was analyzed using a modular compact oscillatory rheometer (MCR 300, Anton Paar, Stuttgart, Germany). The rheometer was equipped with a coaxial cylinder (CC-27, cup diameter: 28.92 mm, bob diameter: 26.66 mm). Flow curves were recorded by increasing the shear rate of the coaxial cylinder system equilibrated to 25 °C logarithmically from 10 $s^{-1}$ to 120 $s^{-1}$. Viscosity was calculated as the slope of the shear stress vs. strain curve and plotted as a function of the shear rate.

*Quantification of ferulic acid equivalents in sugar beet pectin.* Phenolic compounds of sugar beet pectin (citrus pectin and apple pectin were used as references) expressed as ferulic acid equivalents were quantified using a modified method of *Yapo et al.* (2007), whereas the HPLC program was performed as follows. A Gynkotek HPLC-system (Gynkotek, Germering, Germany) equipped with a M480 pump, autosampler Gina 50, degasser (DG 1310 S), connected to a diode array detector (UVD 320) and the Gynkosoft chromatography data system (version 5.50) was used for the chemical analysis. The ferulic acid equivalent was seperated on a column TSK-gel® ODS-80TM 250-4.6 mm, 5 µm (Tosoh Bioscience, Stuttgart, Germany) and a guard column (Supelguard™ LC-18-DB, Supelco, USA). The mobile phase consisted of eluent A: 2% (v/v) acetic acid and eluent B: acetonitrile. Ferulic acid was separated with a gradient program of 1 ml/min with 85% A and 10% B from 0 to 15 min; 50% B and 50% A from 15 to 35 min, 15% A and 85% B from 35 to 36 min and 85% A and 15% B from 36 to 39 min. The injection volume was 50 µl. For the regeneration of the HPLC–column a mobile phase of 75% B was flushed through the column for 4 min. The column was then allowed to equilibrate for 3 mins with the starting conditions. UV-detection was carried out at 325 nm. The peak of ferulic acid was identified by comparing the retention times and UV-spectra with standards. Quantification was performed using a calibration curve

(0, 1, 5, 10, 50, 100 mg/l ferulic acid) obtained with ferulic acid standards. Hydroxybenzoic acid (2 g/l) was used as an internal standard to verify the method. The recovery of the method was checked by spiking the samples with ferulic acid (100 mg/l).

**Emulsion preparation.** Primary emulsions were prepared by homogenizing 5% (w/v) soybean oil and 95% (w/w) aqueous fish gelatin solution at room temperature. A coarse preemulsion was formed by blending oil and emulsifier solution in a high shear blender (Standard Unit, IKA Werk GmbH, Germany) for 2 min and premixes were then passed through a high pressure homogenizer (M110-EH-30, Microfluidics International Cooperation, Newton, MA) three times at 10000 psi. Secondary emulsions were prepared by mixing the primary emulsion with beet pectin solution and 10 mM citrate buffer (pH 3.5) using a vortex to produce a series of emulsions with different oil droplet and pectin concentrations.

**Treatment of emulsions with laccase.** Secondary emulsions with certain oil droplet concentrations (0.1%, 0.5%, and 1.0% (w/v)) were prepared and then treated with laccase by adding different amounts of enzyme (0 – 50 AU) to the emulsions. All samples were stored for 10 days at room temperature. Particle diameter distribution, ζ-potential, and microstructure of all emulsions were analyzed (see below).

**Particle size determination.** Dynamic light scattering was performed using a dynamic light scattering instrument (Nano ZS, Malvern Instruments, Malvern, UK). Emulsions were diluted to a droplet concentration of approximately 0.005% (w/v) with an appropriate buffer to prevent multiple scattering effects. The foundation of this technique is based on the scattering of light by moving particles due to Brownian motion in a liquid. The movement of the particles is then related to the size of the particles. The instrument reports the mean particle diameter (z-average) and the polydispersity index (PDI) ranging from 0 (monodisperse) to 0.50 (very broad distribution).

**ζ-Potential measurements.** Emulsions were diluted to a droplet concentration of approximately 0.005% (w/v) with an appropriate buffer. Diluted emulsions were then loaded into a cuvette of a particle electrophoresis instrument (Nano ZS, Malvern Instruments, Malvern, UK), and the ζ-potential was determined by measuring the direction and velocity that the droplets moved in the applied electric field. The ζ-potential measurements were reported as the average and standard deviation of measurements made from two freshly prepared samples, with 3 readings made per sample.

**Optical microscopy**. All emulsions samples were gently mixed before analysis using a vortex to ensure emulsion homogeneity. One drop of emulsions was placed on an objective slide and then covered with a cover slip. Light microscopy images were taken with an axial mounted Canon Powershot G10 digital camera (Canon, Tokyo, Japan) mounted on an Axio Scope optical microscope (A1, Carl Zeiss Microimaging GmbH, Göttingen, Germany).

**Chemical analysis of serum phase.** The galacturonic acid content of the serum phase of enzyme-treated, secondary emulsions was quantitatively determined by the colorimetric *m*-hydroxydiphenyl assay (*Blumenkrantz et al.*, 1973; *List et al.*, 1985). The emulsions (3.5 ml) were transferred into a centrifugal filter unit (Regenerated Cellulose, 100000 MWCO, Millipore Carrigtwohill, Ireland) and the filter was placed in a centrifuge (Heraeus Centrifuge Biofuge 28RS with # 3746 8 place fixed rotor, 13500 max rpm, Osterode, Germany). The samples were centrifuged at 5000 x g for 15 min to separate the continuous phase from the oil droplets. Secondary emulsions without enzyme were used as reference. Galacturonic acid monohydrate (AUA) was used for the quantification with following concentrations: 10, 20, 40, 60, and 80 mg/l AUA.

**Emulsion stability.** Primary, secondary and secondary, enzyme-treated emulsions were adjusted with a range of different pHs (3.5 – 10 – 3.5) using 0.1 and 1 M HCl and NaOH. All samples were kept for 1 min after reaching the pH value before transferring (5 ml) into glass test tubes and analyzing mean particle distribution and ζ-potential after 24 h. Pictures of all samples were taken using a digital camera.

**Statistical analysis.** All measurements were repeated at least 3 times using duplicate samples. Means and standard deviations were calculated from these measurements using Excel (Microsoft, Redmond, VA, USA).

## RESULTS AND DISCUSSION

### A. Characterization and crosslinking of base polymer solutions

The influence of pH (3 to 10) on the ζ-potential of fish gelatin and sugar beet pectin solutions was investigated (**Figure 1**). The ζ-potential of fish gelatin solutions changed from positive (+19.1 ± 0.8 mV) to negative (-7.2 ± 0.5 mV) as the pH was increased from 3 to 10. **Figure 1** indicates that the p*I* value of the fish gelatin was 8. This value is appreciably higher than the p*I* value (6) reported by the fish gelatin supplier. The high p*I* value may be the result of the acidic pre-treatment of the gelatin. The ζ-potential had a maximum at pH 3 (+19.1 ± 0.8 mV) and a minimum at pH 10 (-7.2 ± 0.5 mV), a fact that is of importance for the subsequently produced primary, gelatin-stabilized base emulsions. In general, protein-stabilized emulsions are sensitive to pH changes since the degree of electrostatic repulsion depends on the magnitude of surface charges (*McClements*, 2004). The droplets tend to aggregate or flocculate at pH values close to the isoelectric point of the adsorbed protein. Therefore, we selected for the preparation of fish gelatin-coated droplets a pH value that was substantially below the p*I* value since the magnitude of the electrical charge of fish gelatin was higher. Aside from the high p*I*, marine gelatin is not associated with the risk of outbreaks of Bovine Spongiform Encephalopathy and can be used for the production of goods sold in Islamic, Judaic and Hindu cultures (*Karim et al.*, 2009). Moreover, the positive charge of fish gelatin allowed a subsequent absorption of an anionic polysaccharide for the preparation of a double-layered emulsion.

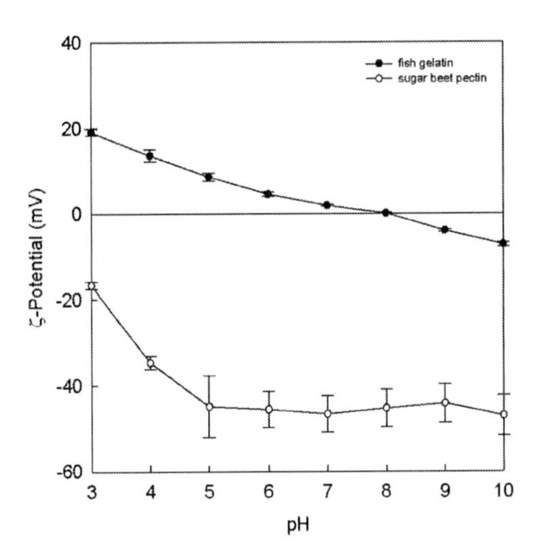


**Figure 1 Influence of pH on electrical charge (ζ-potential) of fish gelatin and sugar beet pectin solutions (10 mM citrate buffer).**

Sugar beet pectin displayed the characteristic behaviour of an anionic polyelectrolyte. The ζ-potential of sugar beet pectin solutions remained negative as the pH was increased from 3 to 10 (**Figure 1**) with magnitudes of charges being lower at lower pH. Sugar beet pectin contains high amounts of arabinose and ferulic acid groups attached to the galactose and arabinose side chains that are more or less deprotonated depending on pH (*Ralet et al.*, 2005). Of the ferulic groups about 50 – 60% are linked to arabinose residues, and 40 – 50% to galactose residues (*Colquhoun et al.*, 1994). Ferulic acids groups may be oxidized by laccase to covalently crosslink pectin molecules via radical reactions (*Yaropolov et al.*, 1994; *Micard et al.*, 1999; *Minussi et al.*, 2002; *Claus*, 2004).

The HPLC-chromatogram of sugar beet pectin is shown in **Figure 2**. For identification of ferulic acid in the pectin sample the retention time of a standard (*trans*-ferulic acid) was used. Ferulic acid was detectable after 11 min. 40 sec at a wavelength of 325 nm. Sugar beet pectin possessed the highest content of ferulic acid with 0.75 ± 0.02%, whereas citrus or apple pectin showed no detectable amounts of ferulic acid (**Figure 2**) as already described by other working groups (*Colquhoun et al.*, 1994; *Buchholt et al.*, 2004; *Ralet et al.*, 2005). We used hydroxybenzoic acid as an internal standard to verify the method.

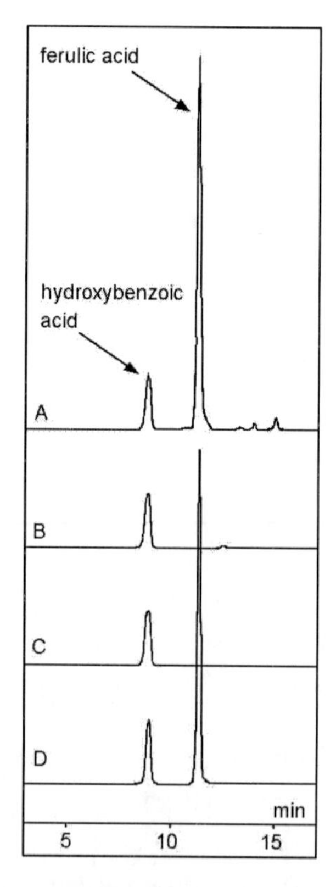


**Figure 2 HPLC-chromatogram of sugar beet pectin (A), apple pectin (B), citrus pectin (C), and *trans*-ferulic acid as a standard (D) (hydroxybenzoic acid was used as internal standard).**

**Determination of laccase-induced crosslinking of sugar beet pectin**

Effect of addition of laccase to sugar beet pectin solutions was investigated according to *Littoz et al.* (2008). Sugar beet pectin showed a peak at 325 nm (pH 3.5) in the UV absorption spectrum (300 – 800 nm). According to *Synytsya et al.* (2003) this peak may be attributed to ferulic acid. The height of absorption should therefore decrease if an enzymatic oxidation of ferulic acid takes place, a fact that could be used to monitor the enzymatic oxidation. Laccase was added in different amounts (0 – 50 AU) to sugar beet pectin solutions (0.1% (w/w) and UV absorption at 325 nm measured. The sugar beet pectin had a high absorbance at 325 in the absence of laccase. After adding laccase in different amounts to the pectin solutions the absorbance decreased over time indicating a crosslinking between pectin molecules (**Figure 3A**). In contrast to sugar beet pectin, fish gelatin did not have any characteristic absorption peak at 325 nm and did not change the range between 250 and 350 nm when laccase was added, suggesting that fish gelatin was not affected by addition of laccase.

A further proof of the pectin-specificity of the reaction of laccase may come from viscosity measurements. We hypothesized that if the reaction is pectin specific, laccase may increase viscosity of pectin solution but not the viscosity of fish gelatin solution since diferulic acid bonds may only be formed between ferulic acid containing pectin molecules. Therefore, the apparent viscosity of pectin and gelatin solutions was determined at shear rates ranging from 10 $s^{-1}$ to 120 $s^{-1}$. After adding different enzyme concentrations to beet pectin solutions (0.5% (w/w), pH 3.5) the viscosity of solutions increased (**Figure 3B**). This indicates that laccase oxidized ferulic acid groups to oxygen-radicals which are reacting further to form diferulic acid bonds between the individual biopolymer molecules. *Kuuva et al.* (2003) and *Norsker et al.* (2000) investigated the rheological properties of laccase-induced sugar beet gels. It was demonstrated that no gelling was observed without laccase, but that laccase had a significant influence on the gelling rate and crosslinking of beet pectin depending on environmental conditions. The apparent viscosity of fish gelatin solutions (0.5% (w/w)) did not increase after enzyme treatment which suggests that no crosslinking occurred between fish gelatin molecules (**Figure 3B**). Results of the UV-visible and rheological investigations therefore indicate a laccase-induced crosslinking of only sugar beet pectin under prevalent conditions.

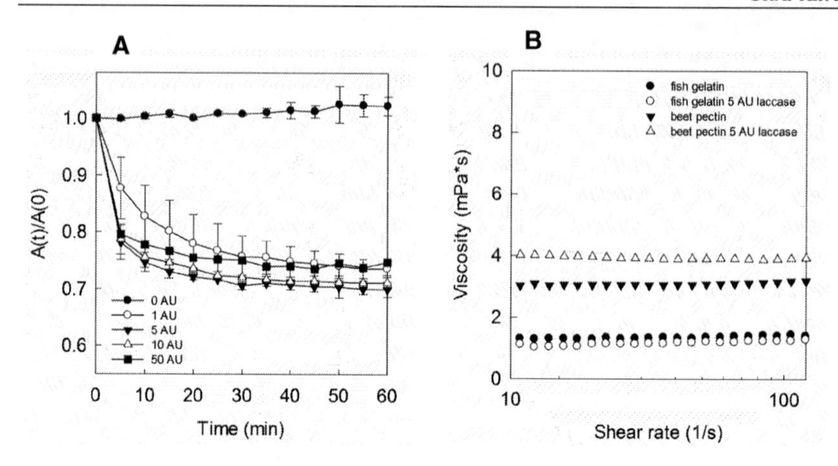


**Figure 3 Influence of enzyme concentration on (A) time-dependent of the UV-visible absorbance at 325 nm, and (B) viscosity of sugar beet pectin and fish gelatin solutions (0.5% (w/w), citrate buffer 10 mM, pH 3.5) as a function of shear rate.**

## B. Multilayering of emulsion droplets

### Properties of fish gelatin-stabilized particles in primary emulsions

We examined the mean particle size distribution and ζ-potential of freshly prepared primary emulsions. While fish gelatin (hydrophilic-lipophilic-balance (HLB) value 9.8

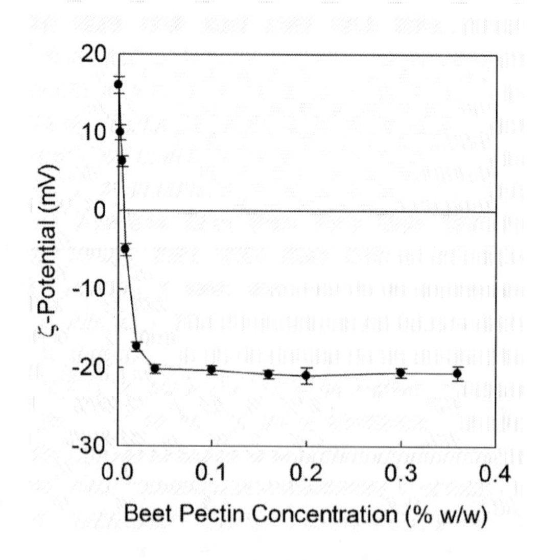


**Figure 4 Particle charge (ζ-potential) of fish gelatin-stabilized emulsions as a function of added sugar beet pectin (0.1% (w/v) soybean oil, 0.019% (w/w) fish gelatin, and 0.04% (w/w) beet pectin, 10 mM citrate buffer, pH 3.5).**

(*Belitz et al.*, 2001)) has been shown to have a lower affinity for oil-in-water interfaces than for example milk proteins, it is nevertheless capable of acting as an emulsifier in oil-in-water emulsions (*Muller et al.*, 1994; *Surh et al.*, 2006). The electrical charge of the fish gelatin-stabilized oil droplets was positive at pH 3.5 because of the high p*I* values of the fish gelatin (**Figure 1** and **4**). The ζ-potential of primary emulsion was +18 ± 2 mV. Dynamic light scattering measurements showed monomodal distributed primary emulsions with a mean particle diameter of 200 ± 20 nm (**Figure 5**). The polydispersity index was 0.076 ± 0.018 indicating a narrow distribution, a fact that is important in preparing multilayered emulsions.

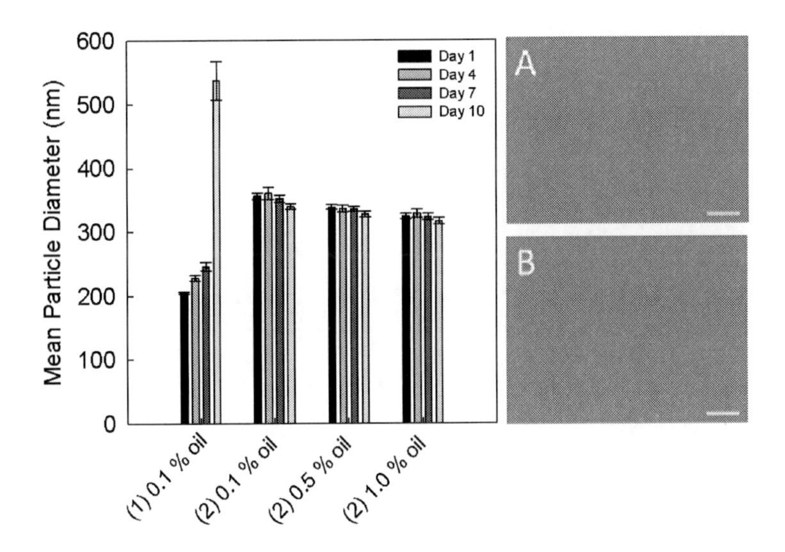


**Figure 5 Influence of storage time on mean particle diameter (z-average) of primary (1) and secondary emulsions (2) containing 0.1 – 1.0% (w/v) oil: (A) primary emulsion and (B) secondary emulsion (scale bar 50 μm)**

Our results are in contrast to those reported by *Surh et al.* (2006) and *Dickinson* (2001) where the presence of a small population of large droplets (d > 10 μm) after homogenization was observed. This was attributed to the low surface activity of fish gelatin compared with globular proteins. Our microscopic images of fish gelatin-stabilized emulsions did not show any large particles or destabilized oil, as shown in **Figure 5**. The authors of the previous studies hypothesized that the molecular weight could be a key factor influencing the ability of gelatin to form stable oil-in-water emulsions besides the protein concentration. In their studies, O/W emulsions stabilized by low molecular weight gelatin (~55 kDa) contained more large droplets and were more prone to coalescence than emulsions made with high molecular weight fish gelatin (~ 120 kDa). According to the manufacturer´s specification the gelatin we used in our studies was a low molecular weight gelatin, but no coalescence over time was observed. This could be due to differences in manufacturing procedures of the emulsion, i.e. we used a Microfluidizer at relatively high pressures of 10.000 psi to obtain very fine disperse base emulsions.

**Influence of beet pectin concentration on droplet characteristics**

The change in the electrical charge of emulsions (0.1% (w/v) oil, 0.019% (w/w) fish gelatin, and 10 mM citrate buffer (pH 3.5) to which different concentrations of pectin (0 – 0.36% (w/w)) had been added was measured **(Figure 4)**. In the absence of pectin, the electrical charge of primary emulsion was +18 ± 2 mV, indicating that the gelatin in emulsion droplets was highly protonated at pH 3.5 **(Figure 4)**. The electrical charge on the droplets became increasingly negative as pectin was added to the emulsion suggesting that the negatively charged pectin molecules adsorbed to the surface of the positively charged oil droplets forming a gelatin – pectin membrane **(Figure 4)**. The $\zeta$-potential became constant at a value of around -20 ± 2 mV when the pectin concentration exceeded about 0.04% (w/v) indicating that the droplets became fully saturated with beet pectin. Based on these results secondary emulsions could be prepared with different oil droplet concentrations (0.1%, 0.5%, 1.0% (w/v)) using 0.04%, 0.2%, and 0.4% (w/w) pectin, respectively.

The mean particle diameter of primary emulsions increased after adsorption of beet pectin from 200 ± 20 nm to 350 ± 50 nm. Dynamic light scattering measurements showed monomodal distributed secondary emulsions with a polydispersity index of 0.101 ± 0.024 (**Figure 5**).

**Storage of primary and secondary emulsions**

Mean particle size distribution, $\zeta$-potential, and creaming behaviour of primary and secondary emulsions were observed over a storage time of 10 days to assess the stability of emulsions.

Primary emulsions were considerably more prone to droplet aggregation than secondary emulsions (**Figure 5**). In comparison to other protein-stabilized emulsions such as emulsions having a whey protein isolate (WPI) membrane, the magnitude of the droplet $\zeta$-potential of fish gelatin covered emulsions is relatively low ($|\zeta| < 30$ mV), and therefore the repulsion between the droplets may not be sufficiently high to overcome attractive droplet interactions e.g. van der Waals and hydrophobic attractions. In particular, WPI-stabilized O/W emulsions are known to remain long-term stable at acidic conditions that are sufficiently below the p*I* of the protein (*Demetriades et al.*, 1997; *Chanamai et al.*, 2002). Emulsions stabilized by a single gelatin-coating were thus stable for 7 days but size increased and formation of a creamed layer could be observed at > 7 days. This correlates with results described by *Dickinson et al.* (2001). These emulsions showed considerable growth in average droplet size $d_{43}$ over the storage period of one week ($d_{43}$ > 21 μm) and were therefore moderately stable to creaming (*Surh et al.*, 2006).

In contrast, the two layer gelatin-beet pectin membrane yielded emulsions that were stable over the entire storage duration. Mean particle diameter and ζ-potentials remained unchanged over a period of 10 days (**Figure 5**). Secondary emulsions have previously been shown to have an improved stability to droplet aggregation because of relatively strong steric repulsion forces that are associated with the two-layered interfacial membrane (*Ogawa et al.*, 2004). Many other studies have already demonstrated that these emulsions are more stable to creaming or environmental stresses such as high mineral content, thermal processing, freezing and drying (*Moreau et al.*, 2003; *Ogawa et al.*, 2004; *Aoki et al.*, 2005). Two-layered emulsions consisting of β-lactoglobulin-pectin membranes were more stable to droplet aggregation and creaming, as shown by *Littoz et al.* (2008). Others demonstrated better stability of multilayered emulsions with β-lactoglobulin-pectin membranes when held at 30 – 90 °C, at high (≥ 100 mM NaCl) salt content, and at pH 3 – 5 (*Guzey et al.*, 2006). Lecithin-chitosan-coated oil droplets formed at pH 3 were stable to aggregation at ≤ 500 mM $CaCl_2$, whereas single-layered lecithin-coated shown aggregation at ≥ 300 mM $CaCl_2$ (*Aoki et al.*, 2005).

The optimum sugar beet pectin concentration required to form stable, non-aggregated oil droplets coated with pectin was 0.04% (w/w), and therefore this concentration was used for all subsequent experiments. Moreover, the above described experiments on the polymer solutions (section A) showed that 5 – 50 AU laccase were sufficient to promote crosslinking of sugar beet pectin solutions. Based on these results, an enzyme treatment of secondary emulsions was applicable in order to determine the laccase-induced influence on multilayered membranes in oil-in-water emulsions.

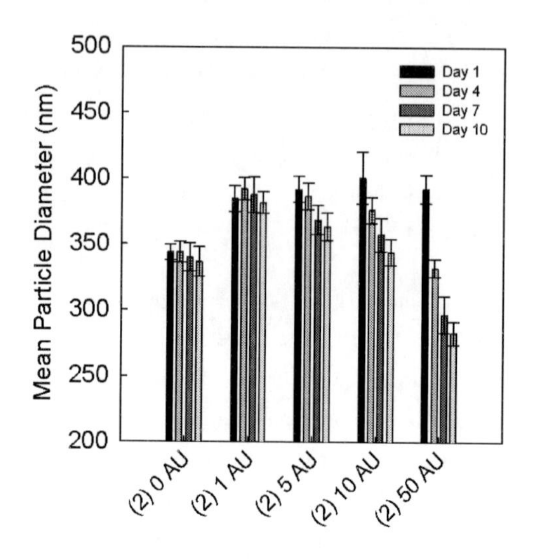


**Figure 6 Changes in mean particle diameter of secondary emulsions upon addition of enzyme (0 – 50 AU) during storage for 10 days.**

## C. Enzyme treatment of emulsions

### Influence of laccase on droplet characteristics over time

The mean particle diameter of laccase-treated, secondary emulsions with different oil droplet concentrations (0.1%, 0.5%, and 1.0% (w/v)) was measured at day 0, 4, 7, and 10 after enzyme addition (**Figure 6**). Microscopic images showed that no crosslinking between single droplets occurred, and only pectin molecules in the membrane layers were crosslinked regardless of the oil droplet concentration (**Figure 7**).

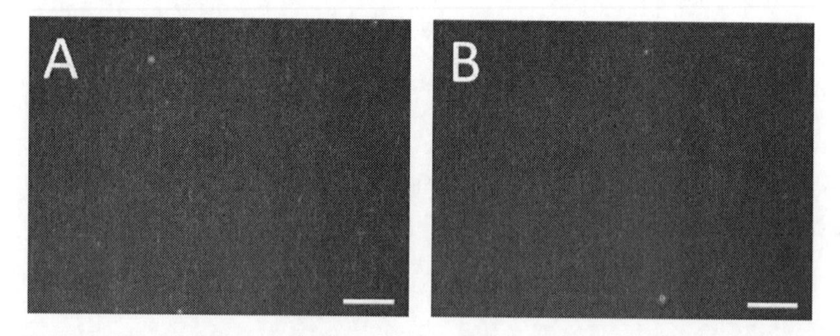


**Figure 7 Photographic images of secondary emulsions before (A) and after (B) enzyme treatment (scale bar 50 μm).**

The mean particle diameter of secondary emulsions increased initially after enzyme addition possibly due to a crosslinking between excess pectin molecules in the surrounding water phase and the pectins adsorbed to the oil droplet surface (**Figure 8**). Interestingly, the mean particle droplet size of all emulsions regardless of oil droplet concentrations after the initial increase decreased over time and enzyme concentration (**Figure 6**). ζ-potential measurements showed that the ζ-potential of the enzyme-treated droplets did not change over time suggesting that pectin remained attached to the surface of the particles (**Figure 9**). Potentially, a structural rearrangement of the pectin molecules caused by crosslinking of ferulic acid could be the reason for the shrinkage of the particles. Pectins usually have an extremely complex composition (*Willats et al.*, 2006). The main structural features include homogalacturonic ("smooth") and rhamnogalacturonic ("hairy") regions (*Thakur et al.*, 1997; *Ralet et al.*, 2005). Homogalacturonan is proposed to be a long side chain of rhamnogalacturonan to which a variety of glycan chains (mainly arabinan and galactan) are attached to (*Vincken et al.*, 2003; *Willats et al.*, 2006). The enzymatic cross-linking of ferulic acid which is mostly esterified to arabinose and galactose residues could assemble the arabinose side chains creating a more compact membrane (**Figure 10**).

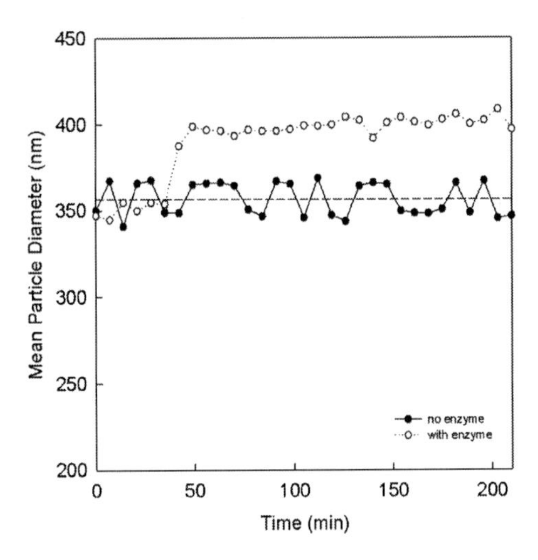


**Figure 8 Initial increase of mean particle diameter of secondary emulsions after enzyme addition.**

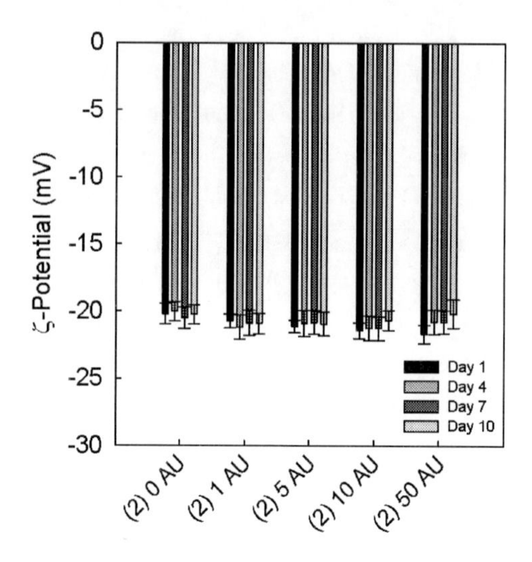


**Figure 9 Changes in ζ-potential of secondary emulsions (0.1% (w/v) oil droplet concentration) after enzyme treatment.**

On the other hand, the decrease in mean particle diameter could be due to an unexpected hydrolytic activity of the laccase itself. Indeed, when we analyzed the serum phase for potential reaction products by separating the serum phase from the oil droplets through centrifugation, an increase of galacturonic acid over time was found (**Table 1**). The increase of absorption at a wavelength of 520 nm indicates a partial hydrolysis of the polygalacturonic backbone catalyzed by the enzyme leading to a decrease of the mean particle diameter over time (**Figure 10**).

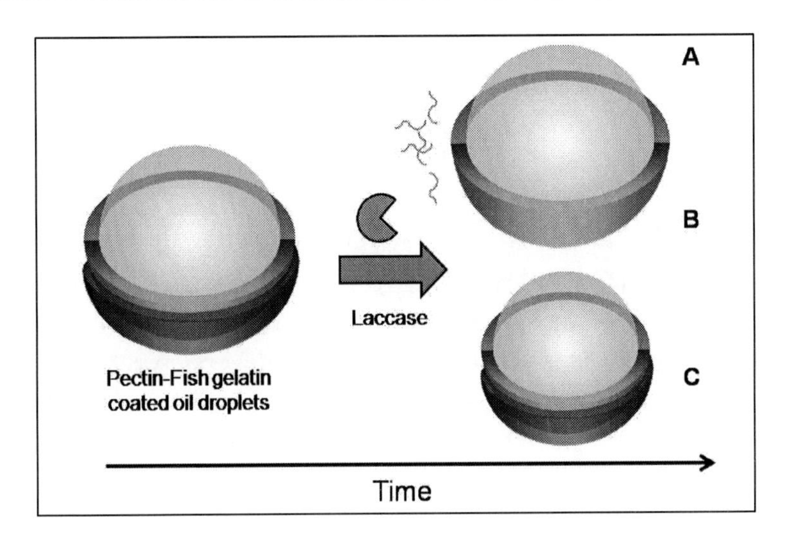


**Figure 10 Schematic representation: origin of shrinkage of secondary, enzyme-treated emulsions (A) desorption of adsorbed pectin, (B) enzymatic hydrolysis of pectins, (C) restructuring of membrane.**

### Effect of enzymatic cross-linking on pH-stability

The effects of different pH on mean particle size and ζ-potential of primary, secondary and enzyme-treated, secondary emulsion was determined (**Figure 11**). We hypothesized that beet pectin that had been adsorbed to the surface of protein-coated oil droplets and then crosslinked would stay on the surface when the pH was shifted from 3.5 to 10 and back to 3.5. In contrast, non cross-linked pectin may detach since charge repulsions between the pectin and gelatin polymers occur at high pH due to deprotonation of the gelatin. Moreover, if crosslinked pectin membranes remained attached, the secondary emulsion would be more stable at high pH values.

**Table 1 Influence of enzyme activity and storage time on hydrolysis of adsorbed beet pectin in fish gelatin-coated oil droplets.**

| Time (days) | No enzyme<br>Uronic acid (µg/g) | With enzyme<br>Uronic acid (µg/g) |
|---|---|---|
| 1 | 442 | 765 |
| 2 | 356 | 851 |
| 3 | 441 | 1174 |
| 4 | 509 | 1631 |
| 8 | 499 | 1960 |
| 10 | 442 | 2066 |

The primary emulsion had a positive charge (+20 mV) at pH 3.5. When the primary emulsion was adjusted to pH 10 the oil droplets became negatively charged (-44 mV). At a pH of 5, the droplets started to strongly aggregate because this pH was close to the p*I* of the gelatin and thus repulsive forces were insufficiently high to prevent droplet-droplet aggregation and collisions. Thus, droplet flocculation could be observed at pH 5 (z-average particle diameter > 600 nm). When the primary emulsion was then adjusted back to pH 3.5 the droplet charge became positive again (+20 mV) (**Figure 11**). The droplets did not remain aggregated at pH 5 after the pH was adjusted to 10 and back to 3.5 which indicated that the droplet flocculation at pH 5 was reversible within the experimental time scale (z-average diameter 220 nm), a fact that has been previously described by *Chanamai et al.* (2002) and *Onsaard et al.* (2006). They demonstrated in case of WPI-stabilized emulsions that the emulsions tend to aggregate significantly at pH 4 - 5 which is close to the p*I* of the protein, but no flocculation was observed at pH values above or below the p*I*. Additionally, *Littoz et al.* (2008) showed that droplet aggregation of β-lactoglobulin-coated oil droplets were partially reversible when the pH was adjusted from pH 4.5 to 7.0.

In the secondary emulsion, the ζ-potential of the droplets changed form -20 mV to -38 mV as the pH was increased from 3.5 to 10 (**Figure 11**). Thus, at the highest pH value of 10 the pectin did not remain attached to the surface of the primary emulsion droplets. This phenomenon could be attributed to the fact that the surface of the gelatin-coated droplets is negatively charged. The electrostatic repulsion between the absorbed gelatin molecules and the pectin polymers causes them to detach from the droplet surface.

Measurements of the mean particle diameter (z-average) of the secondary emulsion depending on pH showed that there was a decrease in particle size as the pH increases (**Figure 11**). This is a further proof that the pectin molecules detached from the gelatin-coated oil droplets. Finally when the pH of laccase-treated secondary emulsion was increased from pH 3.5 to 10 and then decreased back to 3.5 the droplets became increasingly negative charged but the mean particle size remained constant (**Figure 11**) suggesting that laccase had successfully linked the pectin molecules together preventing a detachment.

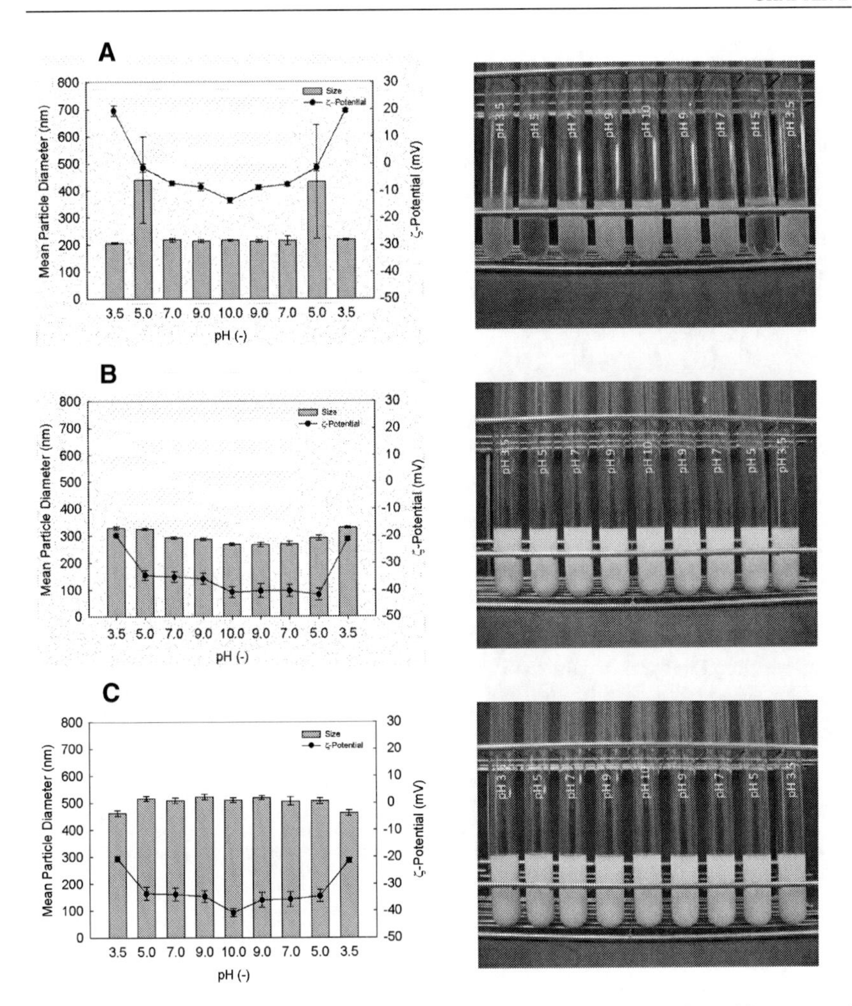


**Figure 11 Mean particle diameter and photographic images of primary (A), secondary (B), and laccase-treated secondary emulsion (C).**

When the pH was adjusted back to 3.5 the pectin stayed attached to the surface of the droplets. Additional storage tests at pH 10 showed no changes in mean particle diameter of enzyme-treated, secondary emulsions, a further proof for the distinct stabilization of crosslinked multilayered emulsions (**Figure 12**).

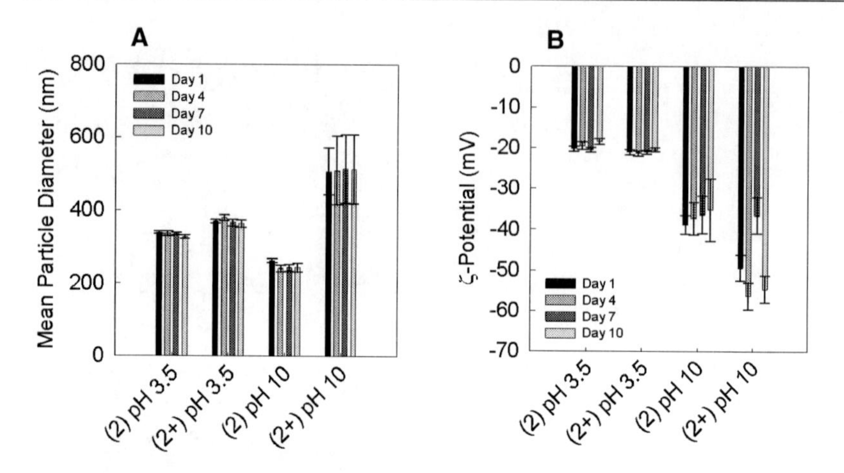


**Figure 12 Mean particle diameter (A) and z-potential (B) of secondary (2) and laccase-treated, secondary emulsions (2+) as a function of time under alkaline environmental conditions (pH 10).**

## Mechanistic explanation

Our results have shown that an enzymatic-induced crosslinking of biopolymers in multilayered oil-in-water emulsions is possible and improves the pH-stability of these emulsions. In summary, we would like to highlight some key insights obtained by our study, as illustrated in **Figure 13**:

- Fish gelatin may be used as an emulsifier to form stable oil-in-water emulsions under acidic conditions with a single membrane layer. The emulsions are prone to time- and pH-dependent aggregation and creaming.

- The formation of multilayered emulsions using layer-by-layer electrostatic deposition method increases storage- and pH-stability of emulsions. However, a pH-triggered detachment of sugar beet pectin under alkaline conditions may be observed and leads to a dissociation of the membrane complex.

- Laccase promotes an enzymatic crosslinking of adsorbed sugar beet pectin in multilayered emulsions improving stability of emulsions over a broad pH range. Laccase-treated emulsions are able to withstand in alkaline environmental conditions for a period of time (**Figure 12**). Laccase did not affect fish gelatin-stabilized emulsions.

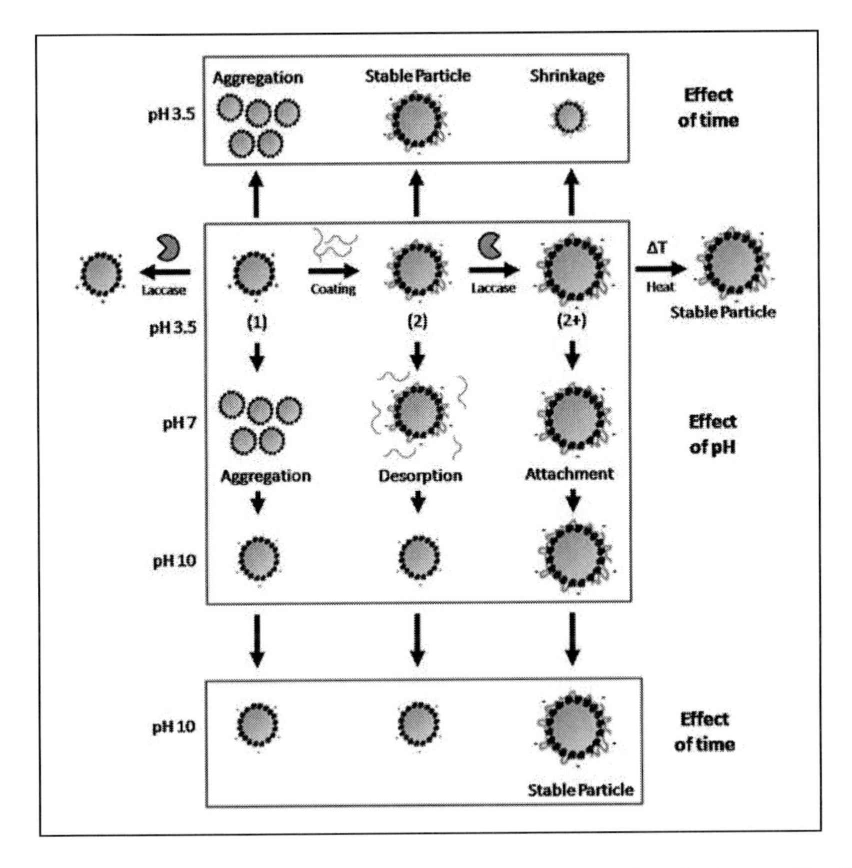


**Figure 13 Schematic mechanism of laccase-induced crosslinking in 2-layered emulsions: (1) primary emulsion, (2) secondary emulsion, (2+) laccase-treated, secondary emulsion.**

- A laccase-induced aggregation of single particles stabilized by multilayered membranes was not observed but mean particle diameter of emulsions initially increases after enzyme addition.

- A subsequent heating step (> 60 °C) of laccase-treated emulsions could inactivate the enzyme and therefore prevent an enzymatic-induced partial hydrolysis of beet pectin membranes to from stable particles (**Figure 14**).

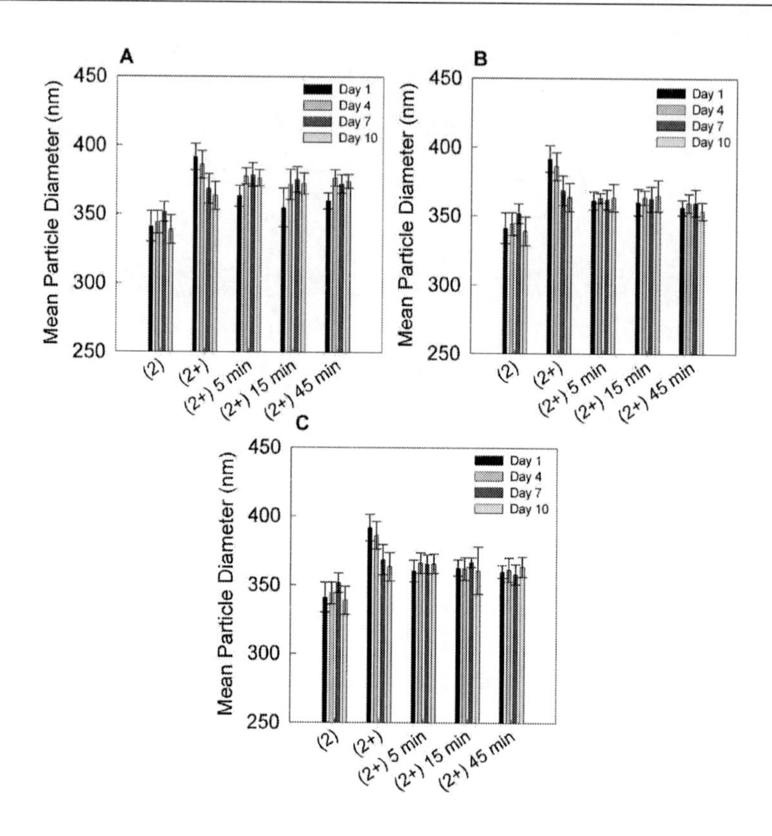


**Figure 14 Effect of heat treatment on mean particle diameter of secondary, crosslinked emulsions: (A) 30 °C, (B) 60 °C, and (C) 85 °C in comparison to unheated emulsions.**

## CONCLUSIONS

This study has demonstrated that laccase could be a useful tool to covalently crosslink biopolymers in multilayered emulsions. The results have shown that pectin remains attached to the surface after increasing the pH from 3.5 to 10 and back. Furthermore, an enzymatic treatment of protein-polysaccharide stabilized particles could form tight membranes resisting harsh environmental conditions and therefore widen the functionality if such systems were used as carriers of functional ingredients. An enzymatic crosslinking of membranes in delivery systems may be robust enough for general applications in which alterations to ionic strength, heating, drying, and freeze-thawing are common in place. Nevertheless, the influence of membrane thickness on the ability of enzymes to crosslink the interfacial biopolymers remain to be investigated. The accessibility of the enzyme to the substrate may play key role. For example, the efficiency of the reaction could decrease if polymers are "buried" under other polymers in a multilayer membrane or if other steric hindrance factors are at play. To date, relatively little is known as to impact of the spatial distribution of molecules on enzymatic crosslinking. Furthermore, future investigations should also focus on the effect of droplet concentration. For example, a crosslinking of membranes in a tightly packed system could lead to inter- rather than intra-crosslinking thereby creating particulate gel networks, with substantially different rheological and functional properties.

## ACKNOWLEDGEMENTS

We would like to thank Herbstreith & Fox KG (Neuenbürg, Germany) for generously providing us with pectin samples.

## REFERENCES


Aoki, T., Decker, E. A., et al. (2005): Influence of environmental stresses on stability of O/W emulsions containing droplets stabilized by multilayered membranes produced by a layer-by-layer electrostatic deposition technique. Food Hydrocolloids 19(2): 209-220.

Belitz, H.-D., Grosch, W., et al. (2001): Lehrbuch der Lebensmittelchemie. Berlin, Springer-Verlag GmbH.

Blumenkrantz, N., Asboe Hansen, G. (1973): New method for quantitative determination of uronic acids. Analytical Biochemistry 54(2): 484-489.

Buchholt, H. C., Christensen, T. M. I. E., et al. (2004): Preparation and properties of enzymatically and chemically modified sugar beet pectins. Carbohydrate Polymers 58(2): 149-161.

Chanamai, R., McClements, D. J. (2002): Comparison of gum arabic, modified starch, and whey protein isolate as emulsifiers: Influence of pH, $CaCl_2$ and temperature. Journal of Food Science 67(1): 120-125.

Claus, H. (2004): Laccases: structure, reactions, distribution. Micron 35(1-2): 93-96.

Colquhoun, I. J., Ralet, M.-C., et al. (1994): Structure identification of feruloylated oligosaccharides from sugar-beet pulp by NMR spectroscopy. Carbohydrate Research 263(2): 243-256.

Dalgleish, D. G. (2006): Food emulsions - their structures and structure-forming properties. Food Hydrocolloids 20(4): 415-422.

Demetriades, K., Coupland, J. N., et al. (1997): Physical properties of whey protein stabilized emulsions as related to pH and NaCl. Journal of Food Science 62(2): 342-347.

Dickinson, E., Lopez, G. (2001): Comparison of the emulsifying properties of fish gelatin and commercial milk proteins. Journal of Food Science 66(1): 118-123.

Færgemand, M., Otte, J., et al. (1998): Cross-Linking of whey proteins by enzymatic oxidation. Journal of Agricultural and Food Chemistry 46(4): 1326-1333.

Gu, Y. S., Decker, A. E., et al. (2005): Production and characterization of oil-in-water emulsions containing droplets stabilized by multilayer membranes consisting of β-lactoglobulin, ι-carrageenan and gelatin. Langmuir 21(13): 5752-5760.

Gu, Y. S., Decker, E. A., et al. (2004): Influence of pH and ι-carrageenan concentration on physicochemical properties and stability of β-lactoglobulin-stabilized oil-in-water emulsions. Journal of Agricultural and Food Chemistry 52(11): 3626-3632.

Gübitz, G. M., Paulo, A. C. (2003): New substrates for reliable enzymes: Enzymatic modification of polymers. Current Opinion in Biotechnology 14(6): 577-582.

Guzey, D., McClements, D. (2006): Influence of environmental stresses on o/w emulsions stabilized by β-lactoglobulin–pectin and β-lactoglobulin–pectin–chitosan membranes produced by the electrostatic layer-by-layer deposition technique. Food Biophysics 1(1): 30-40.

Guzey, D., McClements, D. J. (2006): Formation, stability and properties of multilayer emulsions for application in the food industry. Advances in Colloid and Interface Science 128-130: 227-248.

Hernàndez-Balaba, E., Taylor, M. M., et al. (2009): Properties of biopolymers produced by transglutaminase treatment of whey protein isolate and gelatin. Bioresource Technology 100(14): 3638-3643.

Karim, A. A., Bhat, R. (2009): Fish gelatin: Properties, challenges, and prospects as an alternative to mammalian gelatins. Food Hydrocolloids 23(3): 563-576.

Kuuva, T., Lantto, R., et al. (2003): Rheological properties of laccase-induced sugar beet pectin gels. Food Hydrocolloids 17(5): 679-684.

List, D., Buddruß, S., et al. (1985): Determination of pectin with metaphenylphenol. Pectinbestimmung mit meta-Phenylphenol 180(1): 48-52.

Littoz, F., McClements, D. J. (2008): Bio-mimetic approach to improving emulsion stability: Cross-linking adsorbed beet pectin layers using laccase. Food Hydrocolloids 22(7): 1203-1211.

Ma, H., Forssell, P., et al. (2011): Improving laccase catalyzed cross-linking of whey protein isolate and their application as emulsifiers. Journal of Agricultural and Food Chemistry 59(4): 1406-1414.

Matheis, G., Whitaker, J. R. (1987): A review: Enzymatic cross-linking of proteins applicable to foods. J. Food Biochem. 11(4): 309-327.

Mattinen, M.-L., Hellman, M., et al. (2008): Laccase and tyrosinase catalyzed polymerization of proteins and peptides. Journal of Biotechnology 136(Supplement 1): S318-S318.

Mattinen, M.-L., Kruus, K., et al. (2005): Laccase-catalyzed polymerization of tyrosine-containing peptides. FEBS Journal 272(14): 3640-3650.

McClements, D. J. (2004): Food emulsions: Principles, practice, and techniques. Boca Raton, CRC Press.

McClements, D. J. (2004): Protein-stabilized emulsions. Current Opinion in Colloid & Interface Science 9(5): 305-313.

Mezzenga, R., Ulrich, S. (2010): Spray-dried oil powder with ultrahigh oil content. Langmuir 26(22): 16658-16661.

Micard, V., Thibault, J. F. (1999): Oxidative gelation of sugar-beet pectins: Use of laccases and hydration properties of the cross-linked pectins. Carbohydrate Polymers 39(3): 265-273.

Minussi, R. C., Pastore, G. M., et al. (2002): Potential applications of laccase in the food industry. Trends in Food Science & Technology 13(6-7): 205-216.

Moreau, L., Kim, H. J., et al. (2003): Production and characterization of oil-in-water emulsions containing droplets stabilized by β-lactoglobulin-pectin membranes. Journal of Agricultural and Food Chemistry 51(22): 6612-6617.

Muller, H. J., Hermel, H. (1994): On the relation between the molecular mass distribution of gelatin and its ability to stabilize emulsions. Colloid & Polymer Science 272(4): 433-439.

Norsker, M., Jensen, M., et al. (2000): Enzymatic gelation of sugar beet pectin in food products. Food Hydrocolloids 14(3): 237-243.

O'Connell, J. E., de Kruif, C. G. (2003): β-Casein micelles; cross-linking with transglutaminase. Colloids and Surfaces A: Physicochemical and Engineering Aspects 216(1-3): 75-81.

Ogawa, S., Decker, E. A., et al. (2004): Production and characterization of O/W emulsions containing droplets stabilized by lecithin-chitosan-pectin multilayered membranes. Journal of Agricultural and Food Chemistry 52(11): 3595-3600.

Onsaard, E., Vittayanont, M., et al. (2006): Comparison of properties of oil-in-water emulsions stabilized by coconut cream proteins with those stabilized by whey protein isolate. Food Research International 39(1): 78-86.

Ralet, M. C., André-Leroux, G., et al. (2005): Sugar beet (Beta vulgaris) pectins are covalently cross-linked through diferulic bridges in the cell wall. Phytochemistry 66(24): 2800-2814.

Riva, S. (2006): Laccases: Blue enzymes for green chemistry. Trends in Biotechnology 24(5): 219-226.

Romoscanu, A. I., Mezzenga, R. (2005): Crosslinking and rheological characterization of adsorbed protein layers at the oil-water interface. Langmuir 21(21): 9689-9697.

Romoscanu, A. I., Mezzenga, R. (2006): Emulsion-templated fully reversible protein-in-oil gels. Langmuir 22(18): 7812-7818.

Schmitt, C., Sanchez, C., et al. (1998): Structure and technofunctional properties of protein-polysaccharide complexes: A review. Critical Reviews in Food Science and Nutrition 38(8): 689-753.

Smiddy, M. A., Martin, J. E. G. H., et al. (2006): Stability of casein micelles cross-linked by transglutaminase. Journal of Dairy Science 89(6): 1906-1914.

Surh, J., Decker, E. A., et al. (2006): Properties and stability of oil-in-water emulsions stabilized by fish gelatin. Food Hydrocolloids 20(5): 596-606.

Synytsya, A., Copiková, J., et al. (2003): Spectroscopic estimation of feruloyl groups in sugar beet pulp and pectin. International Sugar Journal 105(1258): 481-488.

Tanimoto, S. Y., Kinsella, J. E. (2002): Enzymic modification of proteins: Effects of transglutaminase cross-linking on some physical properties of β-lactoglobulin. Journal of Agricultural and Food Chemistry 36(2): 281-285.

Thakur, B. R., Singh, R. K., et al. (1997): Chemistry and Uses of Pectin - A Review. Critical Reviews in Food Science and Nutrition 37(1): 47-73.

Tornberg, E., Hermansson, A. M. (1977): Functional characterization of protein stabilized emulsions: Effect of processing. J. Food Sci. 42: 468-472.

Uyama, H., Kobayashi, S. (2002): Enzyme-catalyzed polymerization to functional polymers. Journal of Molecular Catalysis B: Enzymatic 19-20: 117-127.

Vincken, J. P., Schols, H. A., et al. (2003): If homogalacturonan were a side chain of rhamnogalacturonan I. Implications for cell wall architecture. Plant Physiology 132(4): 1781-1789.

Willats, W. G. T., Knox, J. P., et al. (2006): Pectin: New insights into an old polymer are starting to gel. Trends in Food Science & Technology 17(3): 97-104.

Yapo, B. M., Robert, C., et al. (2007): Effect of extraction conditions on the yield, purity and surface properties of sugar beet pulp pectin extracts. Food Chemistry 100(4): 1356-1364.

Yaropolov, A. I., Skorobogat'ko, O. V., et al. (1994): Laccase - Properties, catalytic mechanism, and applicability. Applied Biochemistry and Biotechnology 49(3): 257-280.

# CHAPTER 3

## Crosslinking of interfacial layers affects the salt and temperature stability of multilayered emulsions consisting of fish gelatin and sugar beet pectin

*Benjamin Zeeb[1], Lutz Fischer[2], Jochen Weiss[1]*

[1] Department of Food Physics and Meat Science, University of Hohenheim, Garbenstrasse 21/25, 70599 Stuttgart, Germany

[2] Department of Food Biotechnology, University of Hohenheim, Garbenstrasse 25,70599 Stuttgart, Germany

**ABSTRACT**

This study assessed the stabilizing effect of enzymatic crosslinking on double-coated emulsions (beet pectin – fish gelatin). The beet pectin layer was crosslinked via ferulic acid groups using laccase (an enzyme that is known to catalyze the oxidation of phenolic groups). Fish gelatin-coated oil droplets (primary emulsion) were mixed at ph 3.5 to promote electrostatic deposition of the beet pectin molecules onto the surfaces of the oil droplets (secondary emulsion). Laccase was then added to promote crosslinking of the adsorbed beet pectin layer. Crosslinked pectin-coated oil droplets had similar or significantly better stability ($p < 0.05$) than oil droplets of primary or secondary emulsions to NaCl addition (0 – 500 mM), $CaCl_2$ addition (0 – 250 mM), and thermal processing (30 – 90 °C for 30 min). Freeze-thaw stability and creaming behaviour of enzyme-treated, secondary emulsions after two cycles (-8 °C for 22 h; +25 °C for 2 h) was significantly improved ($p < 0.05$). These results may have important implications for food manufacturer that are in need of having emulsions with improved physical stability, for example emulsions used in frozen foods for sauces, dips, etc.

**Keywords:** Multilayered emulsions; Gelatin; Sugar beet pectin; Laccase; Environmental stress; Salt stability; Thermal processing; Creaming

## INTRODUCTION

Oil-in-water emulsions are used to prepare a variety of products such as foods, paints, cosmetics and drugs. Emulsions are known to be thermodynamically unstable systems, and hence emulsifiers are needed to create emulsions that are kinetically stable under specific environmental conditions (pH, temperature, storage time) (*McClements*, 2004; *McClements et al.*, 2007). Emulsifiers play a major role during homogenization; for example (i) they decrease the interfacial tension between the oil and water phase and thus reduce the overall free energy to disrupt droplets, and (ii) they form a protective coating around the droplets to prevent droplets from coalescing during a collision (*Dickinson*, 1994; *McClements*, 2004). The manufacturing of emulsions for the food industry requires the use of food-grade ingredients such as food biopolymers (e.g. proteins, polysaccharides). Both proteins and polysaccharides, respectively, may have surface-active properties due to a distinct distribution of hydrophobic and hydrophilic moieties and may be used as emulsifiers in oil-in-water emulsions. In particular, proteins such as soy, caseins, whey proteins, or gelatins are heavily used as food emulsifiers (*Dickinson*, 2003; *McClements*, 2004). Upon adsorption at the oil-water interface, the polymer reorients to maximize contact of hydrophobic groups with the oil phase and maximize contact of the hydrophilic groups with the aqueous phase. The extinction of the polymer membrane into the aqueous phase leads to steric repulsive forces that prevent the droplets from coming into close proximity (*Grigoriev et al.*, 2009). The use of proteins as emulsifiers in food emulsions is, however, limited because their properties vary depending on environmental conditions such as pH, heating, ionic strength, and freezing. This may reduce their ability to prevent droplet aggregation or flocculation. Other studies have previously shown that protein-stabilized emulsions are susceptible to droplet aggregation close to their isoelectric point, at high salt concentrations or when heated to or above certain temperatures (*Demetriades et al.*, 1997; *Demetriades et al.*, 1997; *Gu et al.*, 2004; *Aoki et al.*, 2005). *Chanamai et al.* (2002) demonstrated in their study significant differences in the properties of emulsions stabilized by gum arabic, modified starch or whey protein isolate (WPI). WPI-stabilized oil droplets were highly unstable to aggregation near the p*I* (4.2 to 4.5) of the protein because of a relatively low electrostatic repulsion between the droplets (*Onsaard et al.*, 2006). Heating of emulsions stabilized by WPI caused flocculation in emulsions at temperatures between 70 and 80 °C at pH 7. Moreover, the addition of NaCl or $CaCl_2$ promoted droplet-flocculation at pH 7 (*McClements*, 2004).

The layer-by-layer (LbL) electrostatic deposition technique has been developed to form thicker interfacial membranes consisting of multiple layers of biopolymers to improve stability of emulsions against changes in environmental conditions. The LbL method makes use of the electrostatic attraction of oppositely charged polyelectrolytes to create a multi-composite protective layer (*Guzey et al.*, 2006). To form multilayered membranes surrounding oil droplets, a charge reversal of the surfaces needs to occur. If the polyelectrolyte molecules to be adsorbed have a greater charge density than the droplet surface, then polyelectrolyte monolayers are preferentially formed. In contrast, multilayers may be formed if the charge density of polyelectrolytes is much lower than that of the target droplet surface (*Decher et al.*, 1992; *McClements et al.*, 2009). This process typically involves mixing of a primary, for example, a protein-stabilized emulsion and a polysaccharide solution under conditions at which there is an attractive interaction between the surface of the protein-coated droplets and the oppositely charged polysaccharide. Previously, it has been shown that coating of oil droplets by a protein-polysaccharide complex increased stability to environmental stresses due to changes in interfacial charge, structure, and thickness (*Moreau et al.*, 2003; *Guzey et al.*, 2004; *Iwanaga et al.*, 2008; *Zeeb et al.*, 2012). Specifically, resistance of emulsions to withstand changes in pH, ionic strength, heating, and freezing improved when membranes were reinforced using the LbL deposition technique (*Ogawa et al.*, 2003; *Gu et al.*, 2004; *Aoki et al.*, 2005). For example, *Littoz et al.* (2008) showed that double-layered emulsions consisting of $\beta$-lactoglobulin-pectin membranes were more stable to droplet aggregation and creaming. A number of authors found a better stability of multilayered emulsions to thermal processing (30 – 90 °C), e.g. for emulsions with $\beta$-lactoglobulin-pectin membranes (*Guzey et al.*, 2006), emulsions with SDS-chitosan membranes (*Aoki et al.*, 2005), and emulsions with SDS-fish gelatin membranes (*Ogawa et al.*, 2003). Lecithin-chitosan-coated oil droplets formed at pH 3 were stable to aggregation at $\leq$ 500 mM $CaCl_2$, whereas single layered lecithin-coated emulsions aggregated at $\geq$ 300 mM $CaCl_2$ (*Aoki et al.*, 2005).

Formation of multilayered emulsions requires that an electrostatic attraction between the oppositely charged polyelectrolytes within the interfacial membrane be present; therefore, they have a major limitation in use (*Guzey et al.*, 2006; *Grigoriev et al.*, 2009). Changes in environmental conditions such as pH or ionic strength may decrease the magnitude of electrostatic attraction between the protein-polysaccharide complexes and cause a dissociation of the interfacial membrane. We suggest that a biomimetic approach may be used to overcome this deficiency whereby interfacial membranes are covalently crosslinked using enzymes (*Littoz et al.*, 2008; *Chen et al.*, 2010; *Zeeb et al.*, 2012). Our previous study has shown that laccase may crosslink adsorbed beet pectin in multilayered emulsions, thereby increasing storage and pH-stability at pH 3.5 – 10 (*Zeeb et al.*, 2012). Laccase is a well-studied oxidase that oxidizes polyphenols, methoxy-substituted phenols, and diamines (*Thurston*, 1994; *Claus*, 2004). It has the potential to improve not only the efficiency of production but especially quality attributes of a great variety of food products such as salad dressings, ice creams, coffee whiteners, wine, juices, beers, and baked goods. Application of laccase is considered to be a mild processing technology and can be implemented using conventional processing technologies such as mixing and homogenization steps (*Minussi et al.*, 2002). Thus, little capital investment is required. We also demonstrated in our previous work that a laccase-induced aggregation of single droplets stabilized by complex biopolymer membranes did not occur and that, furthermore, covalently crosslinked multilayered interfacial membranes were able to withstand alkaline environmental conditions for a period of up to 10 days.

The objective of the present study was to compare the stability of primary, secondary, and laccase-treated, secondary emulsions against heating, freeze-thaw cycling, and high salt concentrations. We hypothesize that laccase-treated emulsions may have better resistance against any of the mentioned superimposed stresses than primary and secondary emulsions due to a higher resistance of crosslinked biopolymers against dissociation and mechanical disruption. To this purpose, emulsions were fabricated with a single- or double-layered membrane, treated with laccase and subjected to temperature and salt fluctuations. Their stability was then assessed by recording changes in emulsion properties (particle size and homogeneity of spatial droplet distribution). For our experiments we used fish gelatin as a primary emulsifier due to its high p*I* value which facilitated coating at lower pH with a negatively charged polysaccharide.

## MATERIALS AND METHODS

**Materials.** Sugar beet pectin (#1 09 03 135) was donated by Herbstreith & Fox KG (Neuenbürg, Germany) and used without further purification. As stated by the manufacturer the degree of esterification of the beet pectin was 55%. Cold water fish skin gelatin (#049K0050) was purchased from Sigma-Aldrich Co. (Steinheim, Germany). Its average molecular weight and p*I* value were reported to be ca. 60 kDa and pH 6, respectively. Laccase (#0001437590, from *Trametes versicolor*) was obtained from Sigma-Aldrich Co. (Steinheim, Germany). The laccase obtained was reported to have 20.7 Units per mg (AU) of enzyme. Citric acid monohydrate (#409107294, purity ≥ 99.5%) was obtained from Carl Roth GmbH & Co. KG (Karlsruhe, Germany) and sodium citrate dihydrate (#26996TH, purity ≥ 99.0%) was purchased from SAFC (St. Louis, MO). Octane (#07696APV, purity ≥ 99.0%), decane (#08096LMV, purity ≥ 99.0%), and dodecane (#65796EM, purity ≥ 99.0%) were obtained from Sigma-Aldrich Co. (Steinheim, Germany). Sodium chloride (#3957.1, purity ≥ 99.5%) was purchased from Carl Roth GmbH & Co. KG (Karlsruhe, Germany) and calcium chloride (#K90244218) was obtained from Merck KGaA (Darmstadt, Germany). Analytical grade hydrochloric acid (HCl) and sodium hydroxide (NaOH) were purchased from Carl Roth GmbH & Co. KG (Karlsruhe, Germany). Distilled water was used for the preparation of all samples.

**Solution preparation.** Aqueous emulsifier solutions were prepared by dispersing 1% (w/w) fish gelatin powder into 10 mM citrate buffer (pH 3.5). Sugar beet pectin solutions were prepared by dispersing 2% (w/w) powdered pectin into 10 mM citrate buffer at pH 3.5 followed by stirring overnight to ensure complete hydration. The pH was then adjusted to 3.5 using 0.1 and 1 M HCl and/or 0.1 and 1 M NaOH. Enzyme solutions were prepared by dispersing enzyme powder into 10 mM citrate buffer (pH 3.5) followed by stirring for 30 min.

**Emulsion preparation.** A stock fish gelatin-stabilized emulsion was prepared by mixing 5 g of soybean oil in 95 g fish gelatin solution (1% (w/w) in citrate buffer, 10 mM, pH 3.5) into a glass beaker to obtain a 5% (w/w) primary oil-in-water emulsion at room temperature. The oil and emulsifier solutions were blended using a high shear blender (Standard Unit, IKA Werk GmbH, Germany) for 2 min and were then passed through a high pressure homogenizer (M110-EH-30, Microfluidics International Cooperation, Newton, MA) three times at 10,000 psi (= 68.95 MPa).

The stock emulsion was diluted (1:10) with either 10 mM citrate buffer (pH 3.5) or an aqueous beet pectin solution to prepare a primary emulsion (0% (w/w) beet pectin) or a secondary emulsion (0.2% (w/w) beet pectin) with the same oil droplet content (0.5% (w/w) soybean oil). Laccase-treated secondary emulsions were prepared by mixing the secondary emulsion with citrate buffer (pH 3.5) and the enzyme solution using a vortex. An enzyme/beet pectin ratio of 0.24 mg/4 mg, equivalent to 5 AU, was sufficient to promote interfacial crosslinking of sugar beet pectin (*Zeeb et al.*, 2012).

**Particle size determination.** Mean particle diameter and polydispersity indices were measured using a dynamic light scattering instrument (Nano ZS, Malvern Instruments, Malvern, UK). Emulsions were diluted to a droplet concentration of approximately 0.005% (w/w) with buffer to prevent multiple scattering effects. The technique determines the size of particles from measurements of the Brownian motion of the particles using light scattering. The size is then calculated from the diffusion constant using the Einstein equation (*Dalgleish et al.*, 1995). The instrument reports the mean particle diameter (z-average) and the polydispersity index (PDI) ranging from 0 (monodisperse) to 0.50 (very broad distribution).

**ζ-potential measurements.** Emulsions were diluted to a droplet concentration of approximately 0.005% (w/w) with buffer. Diluted emulsions were then loaded into a cuvette of a particle electrophoresis instrument (Nano ZS, Malvern Instruments, Malvern, UK), and the ζ-potential was determined by measuring the direction and velocity that the droplets moved in the applied electric field. The ζ-potential measurements were reported as the average and standard deviation of measurements made from two freshly prepared samples, with 3 readings made per sample.

**Optical microscopy.** The structures of emulsions were investigated by optical microscopy. All emulsions samples were gently mixed before analysis using a vortex to ensure emulsion homogeneity. One drop of emulsions was placed on an objective slide and then covered with a cover slip. Light microscopy images were taken with an axial mounted Canon Powershot G10 digital camera (Canon, Tokyo, Japan) mounted on an Axio Scope optical microscope (A1 , Carl Zeiss Microimaging GmbH, Göttingen, Germany) at resolution of 40x.

**Emulsion environmental stress tests**. We determined the influence of various kinds of environmental stresses on the mean particle diameter (z-average), ζ-potential and microstructure of primary, secondary and laccase-treated secondary emulsions.

*Salt treatment.* Primary, secondary, and enzyme-treated emulsions were diluted with citrate buffer (10 mM, pH 3.5) to contain the same final oil droplet concentrations (0.5% (w/w)), but different NaCl (0 – 500 mM) or $CaCl_2$ (0 – 250 mM) concentrations. The pH of the samples was adjusted back to 3.5 if any changes due to salt addition occurred. The emulsion samples were stored at room temperature for 24 h before being analyzed.

*Thermal treatment.* Primary, secondary and enzyme-treated emulsions were diluted with citrate buffer (10 mM, pH 3.5) to obtain emulsions with the same final oil droplet concentrations (0.5% (w/w)). Emulsion samples were then transferred into glass test tubes and were incubated in a water bath for 30 min at temperatures ranging from 30 – 90 °C. The samples were then placed immediately into a 25 °C water bath, where they were stored for 24 h prior analysis.

*Combination of thermal and salt treatment.* Primary, secondary and enzyme-treated emulsions were diluted with citrate buffer (10 mM, pH 3.5) to contain a final oil droplet concentration of 0.5% (w/w), but certain NaCl (500 mM) or $CaCl_2$ (250 mM) concentrations. The pH of the samples was again adjusted to 3.5 if any changes after salt addition occurred. Emulsion samples were then transferred into glass test tubes and were incubated in a water bath for 30 min at temperatures ranging from 30 – 90 °C. The samples were then placed into a 25 °C water bath, where they were stored for 24 h prior analysis.

*Freeze-thaw cycling.* Primary, secondary and enzyme-treated emulsions were diluted with citrate buffer (10 mM, pH 3.5) to a final oil droplet concentration of 0.5% (w/w). Emulsions (10 ml) were transferred into glass test tubes and were incubated in a –8 °C salt water bath for 22 h. After incubation, the emulsion samples were thawed by incubating them in a 25 °C water bath for 2 h. This freeze-thaw cycle was repeated two times, and its influence on mean particle diameter, $\zeta$-potential and microstructure was measured after each cycle.

**Accelerated creaming behaviour.** The creaming behaviour of primary, secondary, and laccase-treated secondary emulsions was determined using an accelerated creaming test. Primary, secondary and enzyme-treated emulsions were diluted with citrate buffer (10 mM, pH 3.5) to a final oil droplet concentration of (0.5% (w/w)), and subjected to certain NaCl (500 mM) or $CaCl_2$ (250 mM) concentrations and/or thermal treatments (30 °C, 75 °C, 90 °C each 30 min). Samples of emulsions (2.5 ml) were transferred into a cuvette and the cuvette was placed in a centrifuge (Heraeus Centrifuge Biofuge 28RS with # 3746 8 place fixed rotor, 13500 max rpm, Osterode, Germany). The samples were centrifuged at 2500 x g for 30 min. Oil droplets moved upwards due to centrifugal force, which led to the formation of a clear serum layer at the bottom and a droplet-rich cream layer at the top. Transmission of UV-visible light at 600 nm was measured immediately after centrifugation using a spectrophotometer (HP 8453, Agilent with application software Chemstation Agilent Technologies 95-00, Waldbronn, Germany). The light beam passed through the emulsions at a height of 1 cm from the cuvette bottom, i.e. about 30% of the emulsion's height. An appreciable increase in emulsion transmission was therefore an indication of the fact that the serum layer had risen to at least 30% of the emulsion's height. Citrate buffer was used as reference (transmission = 100%).

**Statistical analysis.** All measurements were repeated at least three times using duplicate samples. Means and standard deviations were calculated from these measurements using Excel (Microsoft, Redmond, VA, USA). Results were analyzed with statistical software SAS 9.2 from SAS institute (Cary, USA). Variance analysis was performed with Tukey test for normal distributed and balanced data ($\alpha = 0.05$). If normal distributed results were doubted, a non-parametric test according to Kruskal-Wallis was performed. For analysis with Kruskal-Wallis test, a Bonferroni correction ($\alpha_{bon} = \frac{k(k-1)}{2}$) was used.

## RESULTS AND DISCUSSION

### Influence of ionic strength on emulsion stability

The ζ-potential of fish gelatin-coated droplets was +20 ± 1 mV regardless of salt concentration or salt type added to the emulsion, a somewhat surprising initial result ($p > 0.05$) (**Figure 1A** and **1B**). The interfacial membranes formed by proteins at low ionic strength are strongly charged at pH values substantially above or below the isoelectric point, and the major mechanism protecting droplets against flocculation at low or high pH values in protein-stabilized emulsions is thus via electrostatic repulsion (*Dickinson*, 1992). Previous studies of protein-stabilized oil-in-water emulsions in the presence of salts demonstrated a reduction of ζ-potential with increasing ionic strength due to electrostatic screening or binding of oppositely charged counterions (*Demetriades et al.*, 1997; *Ogawa et al.*, 2003; *McClements*, 2004; *Grigoriev et al.*, 2009; *Dickinson*, 2010). For example, the ζ-potential of *β*-lactoglobulin coated oil droplets in primary emulsions showed a decreased from +28 mV to +5 mM upon addition of NaCl (0 – 100 mM) (*Guzey et al.*, 2006). *Gu et al.* (2005) described similar effects of NaCl on the stability of WPI-stabilized emulsions. Multivalent counterions such as calcium are more effective at screening than monovalent ions like sodium, a fact that has been demonstrated by *McClements and coauthors* (2004). The ζ-potential of secondary and enzyme-treated, secondary emulsion droplets remained negative for all salt concentrations and salt types (**Figure 1A** and **1B**). However, the magnitude of the ζ-potential decreased with addition of salt, particularly the $CaCl_2$ concentration increased from 0 to 250 mM. A similar result has been previously described by *Surh et al.* (2006, 2007), and may be attributed to the more effective electrostatic screening by the multivalent calcium counter ions (*McClements*, 2004; *McClements*, 2004).

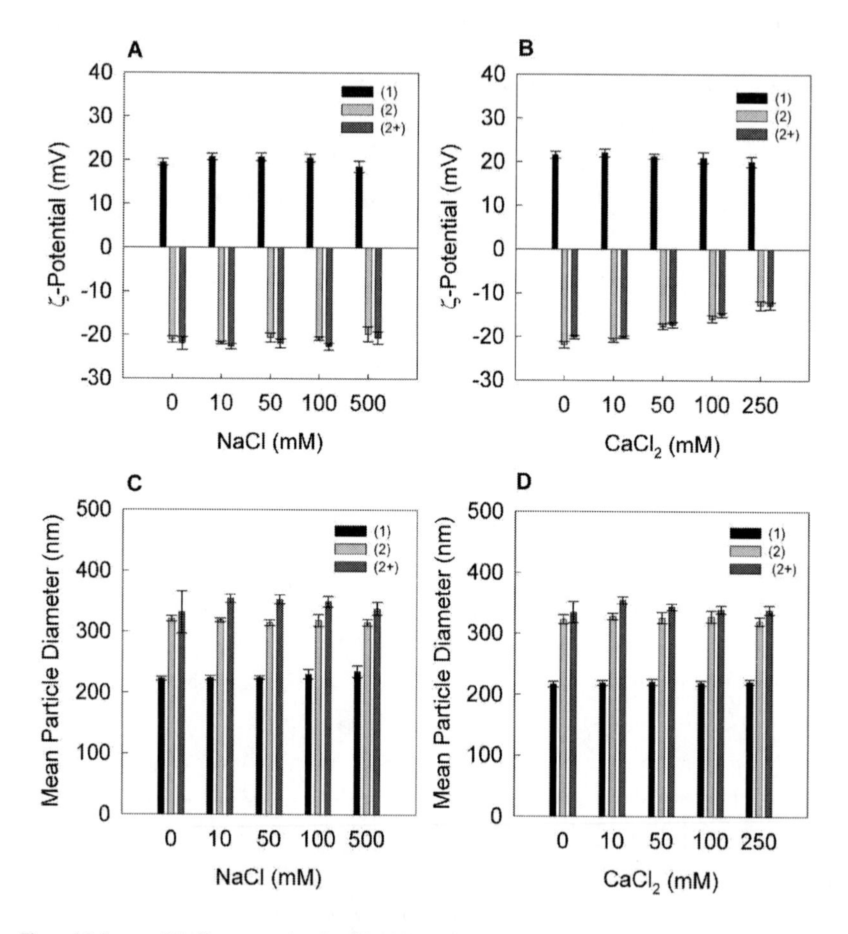


**Figure 1 Influence of NaCl concentration (0 – 500 mM) and $CaCl_2$ concentration (0 – 250 mM) on ζ-potential (A, B) and mean particle diameter (z-average; C, D) of fish gelatin-coated (1), beet pectin-fish gelatin-coated (2), and laccase crosslinked beet pectin-fish gelatin-coated emulsion droplets (2+).**

The observed insensitivity of the ζ-potential to addition of NaCl and $CaCl_2$ may also help to explain the lack of an effect of both salt types on the mean particle diameter of primary, secondary and enzyme-treated emulsion (**Figure 1C** and **1D**). Emulsions with fish gelatin were surprisingly resistant to salt addition. The z-average diameter of fish gelatin-stabilized oil droplets remained constant at 220 ± 20 nm ($p > 0.05$) and did not increase when salt concentrations increased. The emulsions had monomodal droplet size distributions. This suggests that the major mechanism preventing droplet aggregation or flocculation in fish gelatin-stabilized emulsions is steric repulsion, rather than electrostatic repulsion, which is in contrast to milk proteins. To further investigate the unusual salt stability of fish gelatin-stabilized emulsions, we also prepared them at pH 7 (close to p*I* of the fish gelatin) and mixed them with different NaCl (0 – 500 mM) and $CaCl_2$ (0 – 250 mM) concentrations. Although the ζ-potential of oil droplets decreased with increasing salt concentration, no droplet aggregation was observed. Mean particle diameter of fish gelatin-stabilized primary emulsions remained constant (**Figure 2A** and **2B**). These results further support the theory that fish gelatins are remarkably capable steric stabilizer. *Surh et al.* (2007) conducted a study on the influence of pH, emulsifier concentration, and homogenization conditions on the production of stable oil-in-water emulsions stabilized by fish gelatin and found similarly to our results that emulsions were unusually resistant to superimposed stresses. The magnitude of ζ-potential of their fish gelatin-stabilized oil droplets was lower than 20 mV, and because the emulsions showed good creaming stability, author concluded that rather polymeric steric repulsion rather than electrostatic interactions might be involved in stabilizing the emulsions against aggregation or flocculation (*Surh et al.*, 2006).

The mean particle diameter of secondary and enzyme-treated, secondary emulsions remained constant after addition of salt, and all emulsions had monomodal droplet size distributions (**Figure 1C** and **1D**). Apparently, the addition of salt regardless of concentration did not promote droplet aggregation, although the ζ-potential of $CaCl_2$-treated emulsions decreased with increasing salt concentrations, which may be attributed to the effectiveness of charge screening by multivalent ions (*McClements*, 2004). To form secondary emulsions, sugar beet pectin with a degree of esterification of 55% was used. Low-esterified pectin types have been suggested to be more prone to salt-induced flocculation/aggregation since some interchain associations may arise due to ion binding (*Belitz et al.*, 2001). However, the high content of acetyl groups of beet pectin used in our study may reduce salt sensitivity and contribute to the stability of secondary and secondary, enzyme-treated emulsions (*Leroux et al.*, 2003). Generally, multilayered emulsions are much less prone to droplet aggregation in the presence of salt, a fact that a number of studies have previously demonstrated (*Moreau et al.*, 2003; *Ogawa et al.*, 2003; *Aoki et al.*, 2005). For example, lecithin-chitosan-coated oil droplets formed at pH 3 were stable to aggregation at $\leq$ 500 mM $CaCl_2$, whereas single-layered lecithin-coated droplets aggregated at $\geq$ 300 mM $CaCl_2$ (*Aoki et al.*, 2005). Other studies with

NaCl have shown that multilayered interfacial membranes formed with an anionic surfactant and the cationic polysaccharide chitosan have better stabilities than single- surfactant stabilized droplets (*Ogawa et al.*, 2003). The lower salt sensitivity may be attributed to the fact that the range over which the steric repulsion interactions are active increase the thicker the interfacial membranes are, e.g. in secondary emulsions (*McClements*, 2004; *Guzey et al.*, 2006).

The results of our accelerated creaming test are, however, somewhat different in contrast to particle size and ζ-potential measurements and show differences between the performance of the primary, secondary and secondary, enzyme-treated emulsions. The primary emulsion was prone to creaming at high NaCl concentrations (**Figure 3**). The fish gelatin coating apparently did not prevent the droplets in primary emulsions from flocculating and creaming in the centrifugal field. Consequently, an increase in transmission was observed indicative of the serum layer rising to at least 30% of the emulsion's height (*Ogawa et al.*, 2003; *Aoki et al.*, 2005; *Dickinson*, 2010). Similarly, addition of calcium promoted aggregation and creaming. In contrast, secondary and enzyme-treated secondary emulsions did not cream, i.e. no increase in transmission at 600 nm was detected (**Figure 3**). This suggests that either droplets did not aggregate or flocculate or that the density difference between the droplets and the aqueous phase had decreased, preventing the formation of a droplet-rich cream layer. Higher stability to gravitationally or centrifugally induced separation of droplets with higher mass ratios of polymer-containing interfacial layers to oil has recently been described by *McClements et al.* (2011) for single-layered nanoemulsions. Results of the accelerated creaming experiments should, however, be taken with some caution due to the fact that emulsions may behave differently in a centrifugal field than in a gravitational field.

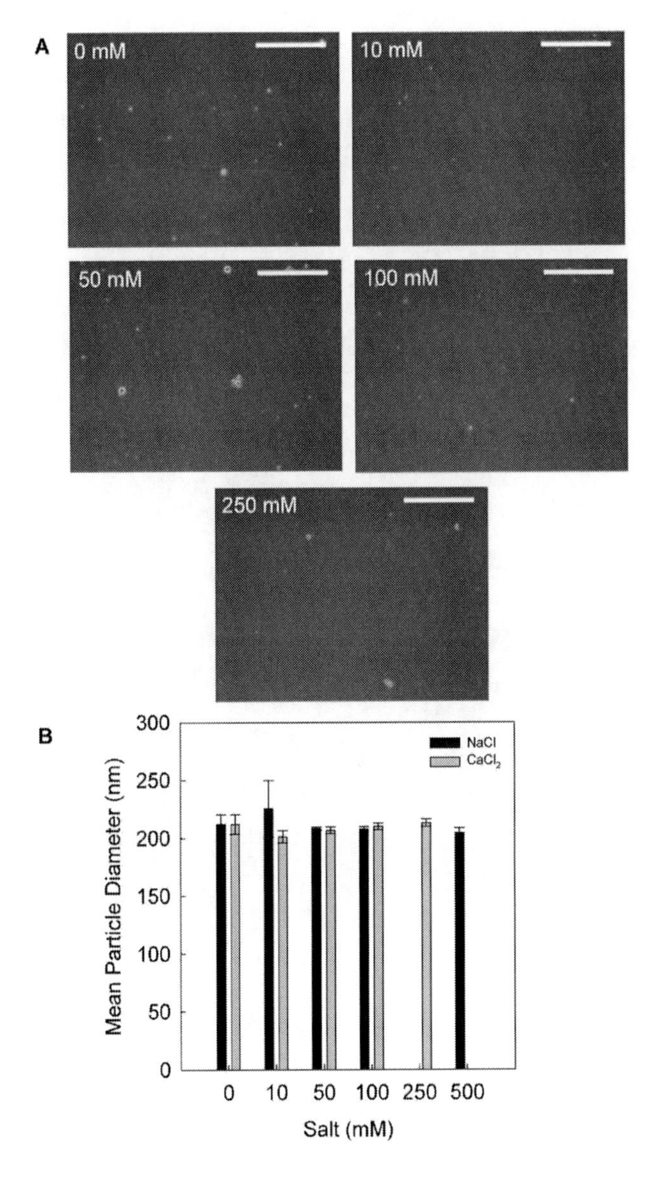


**Figure 2 Influence of $CaCl_2$ concentration (0 – 250 mM) on microstructure (A) and mean particle diameter (B) of fish gelatin-stabilized oil droplets at pH 7 (scale bar 50 μm).**

### Influence of thermal processing on emulsion stability

The influence of heat treatment on the properties and stability of emulsions stabilized by a fish gelatin membrane, a fish gelatin-sugar beet pectin membrane and a fish gelatin-crosslinked sugar beet pectin membrane were studied. Many food emulsions are subject to some kind of heat treatment during their production, storage, or application; for example, many products are pasteurized or sterilized (*McClements*, 2004). The ability to maintain physical stability is thus an important criterion for the manufacture of emulsions.

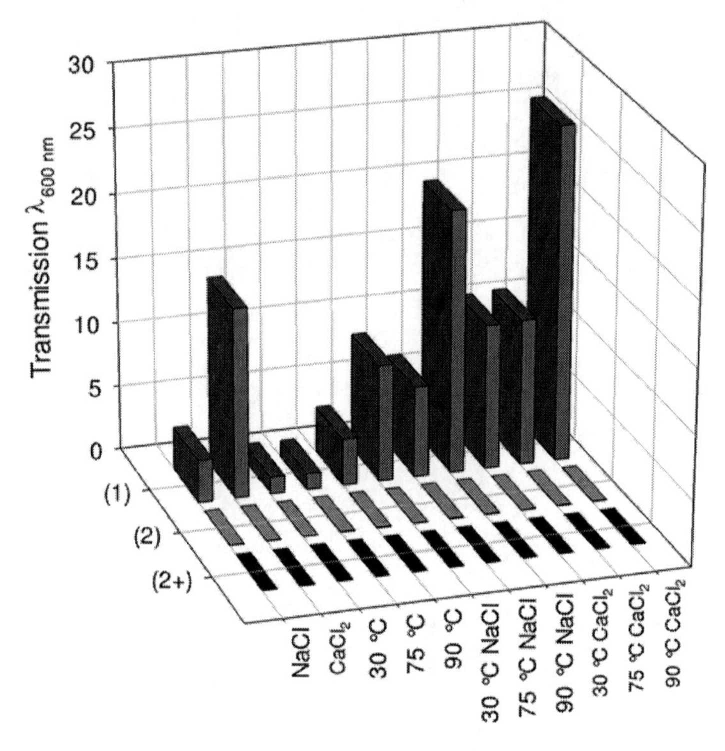


**Figure 3 Influence of salt concentration, salt type, and thermal processing on creaming stability of primary (1), secondary (2), and laccase-crosslinked secondary emulsions (2+).**

There was no significant increase in the mean particle diameter of primary and laccase-treated, secondary emulsions upon isothermal heat treatment ($p < 0.05$), but mean particle diameter of secondary emulsions increased at temperatures exceeding 70 °C ($p < 0.05$) (**Figure 4A**). Nonetheless, no aggregation or coalescence was observed, and the emulsions had monomodal droplet size distributions (PDI 0.103 ± 0.022). Similarly, the ζ-potential of primary, secondary and laccase-treated secondary emulsion samples was not affected to heat treatment (**Figure 4B**). Typically, protein-stabilized emulsions are relatively sensitive to heat treatment because they may undergo a change in conformation when a certain temperature is exceeded (*Demetriades et al.*, 1997). The heat-induced transformation of proteins exposes reactive groups and/or hydrophobic regions from the interior of the folded protein to the aqueous phase thereby promoting droplet-droplet interactions (*McClements*, 2004; *Dickinson*, 2010). Fish gelatin is known to have a hydrophilic random coil structure (*Gu et al.*, 2005). Flexible random-coil polymers such as fish gelatin usually adopt arrangements where the predominantly non-polar regions protrude into the oil phase and the predominantly polar segments protrude into the aqueous phase (*McClements et al.*, 2009). Fish gelatin, in particular, appears to form thin but elastic layers surrounding the oil droplets, thereby stabilizing the oil-in-water emulsions against coalescence and aggregation (*Babel*, 1996; *Avena-Bustillos et al.*, 2006). The high hydroxyproline content in fish gelatin is thought to play an important role in the stabilization of helical chains due to hydrogen bonding to its hydroxyl group, whereas the entire glycin-prolin-hydroxyproline sequence appears to govern the overall thermostability of gelatin (*Burjanadze*, 2000; *Avena-Bustillos et al.*, 2006; *Harrington et al.*, 2009). The high heat stability did, however, not translate into a higher creaming stability; that is, a cream layer was formed in primary emulsions during the accelerated creaming test (**Figure 3**). The increase in the transmission at 30% of the emulsion´s height with increasing treatment temperatures indicates that the susceptibility to creaming increases the higher the emulsion is heated prior to the accelerated creaming test.

For secondary and secondary enzyme-treated emulsions, no creaming was observed (**Figure 3**). This is despite the fact that secondary emulsions not treated with laccase had droplet diameters that were statistically larger than those of laccase-treated secondary emulsions ($p <$ 0.01) (**Figure 4A**). This could be due to a secondary activity of the enzyme promoting the hydrolysis of sugar beet pectin, a fact we observed in our prior study and that has been reported in literature. The nature of the pectin used to coat the droplets governs the thermal stability of emulsions. The pectin backbone consists of α-1,4-linked D-galacturonic acid units (smooth regions), which are interrupted by 1,2-linked L-rhamnose units (hairy regions). The lateral chains consist mainly of D-galactose and L-arabinose; two neutral sugars (*Voragen et al.*, 2008). Sugar beet pectin used in our study differs in their physicochemical properties to other conventional pectins. It has a higher proportion of neutral sugars and a higher content of acetyl groups (*Rombouts et al.*, 1986; *Williams et al.*, 2005). The higher neutral sugar content of sugar beet pectin contributes to an enhanced emulsion stability through viscosity effects, steric hindrance, and electrostatic interactions (*Chanamai et al.*, 2001). The hydrated layer may prevent flocculation of the beet pectin coated oil droplets upon heat treatment.

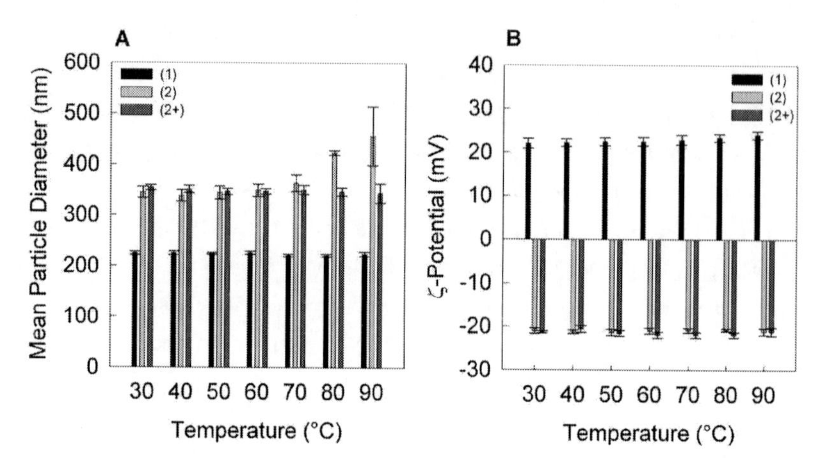


**Figure 4 Impact of isothermal heat treatment on mean particle diameter (A) and ζ-potential (B) of primary (1), secondary (2), and laccase-treated secondary emulsions (2+).**

### Influence of thermal treatment and salt addition on emulsion stability

The mean particle diameter and ζ-potential of oil droplets in primary, secondary, and enzyme-treated secondary emulsion remained constant regardless of combinations of heat and salt treatments. Similarly, optical microscopy images showed that no droplet aggregation occurred (**Figure 5**). Fish gelatin acts as a thermoresistant steric barrier preventing droplet aggregation of primary emulsion (see above). In secondary emulsions, the hydrocolloid beet pectin forms a thick hydrated barrier which is not easily affected by salt addition and heating (*Funami et al.*, 2007; *Morris et al.*, 2010).

Contrary to results of particle size and ζ-potential measurements, accelerated creaming test showed differences between primary, secondary and secondary enzyme-treated emulsions. Fish gelatin-stabilized emulsions were particularly prone to creaming (**Figure 3**). Addition of high amounts of salt and heating to high temperatures led to extensive creaming of primary emulsions. In comparison, secondary and enzyme-treated, secondary emulsions did not cream; that is, no increase in transmission at 600 nm was observed, indicating that the droplets did not aggregate or flocculate or that the density difference decreased preventing the formation of a droplet-rich layer (**Figure 3**). This may be attributed due to the thicker interfacial membrane of secondary and enzyme-treated, secondary emulsion.

### Influence of freeze-thaw cycling on droplet characteristics

Finally, the stability of emulsions to withstand freeze-thaw cycling was investigated. A variety of different physicochemical phenomena occur during frozen storage that may affect the stability and properties of emulsion droplets including fat crystallization, ice formation, freeze concentration, and biopolymer conformational changes (*Guzey et al.*, 2006). When the emulsion is cooled to a temperature at which the fat crystallizes but the water remains liquid, partial coalescence may occur, for example, fat crystals penetrate the membrane and, upon collision, aggregation may occur (*Harada et al.*, 2000; *Grigoriev et al.*, 2009).

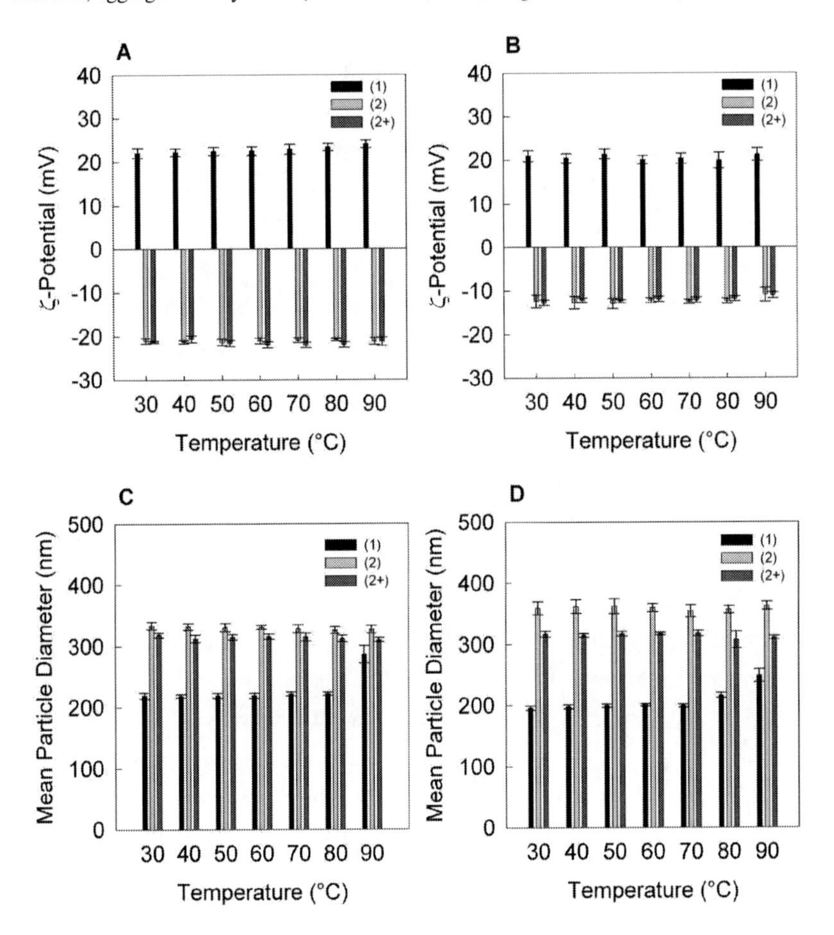


**Figure 5 Impact of isothermal heat treatment, salt concentrations, and salt type on ζ-potential (A, NaCl; B, $CaCl_2$) and mean particle diameter (C, NaCl; D, $CaCl_2$) of primary (1), secondary (2), and laccase-treated secondary emulsions (2+).**

When an emulsion is further cooled to a temperature at which the water starts to crystallize, additional processes are known to occur (*Guzey et al.*, 2006). The crystallization of water pushes the oil droplets into closer proximity because the droplets are restricted to the non-frozen regions and thus concentrate in the aqueous phase (*Saito et al.*, 1999). Second, there may not be sufficient free water present to fully hydrate the droplet surfaces promoting unfavorable droplet-droplet interactions (*Strauss et al.*, 1986; *Ausborn et al.*, 1994; *Komatsu et al.*, 1997). Additionally, the ionic strength of the unfrozen aqueous phase surrounding the droplets is increased due to the formation of ice crystals. This freeze-concentration effect promotes the screening of electrostatic repulsion which may promote droplet aggregation (*Komatsu et al.*, 1997; *Guzey et al.*, 2006). Third, the ice crystals formed during freezing may physically penetrate and disrupt the interfacial membrane making oil droplets more prone to coalescence when the emulsion is thawed. Forth, emulsifiers could adsorb to the surface of ice crystals and thereby cause a decrease in the concentration of emulsifiers in the oil-water interface stabilizing the oil droplets (*Guzey et al.*, 2006). Fifth, the membrane thickness of oil droplets may play a role in the stabilization of emulsions during freezing and thawing in the presence of low molecular weight ingredients (*Cornacchia et al.*, 2011). Finally, the decrease of temperature may cause conformational changes of the adsorbed biopolymers which could lead to a loss of the protective function of the droplet membrane.

In our studies, freeze-thaw cycling caused no changes in ζ-potential of samples, but we did observe differences in emulsions stability of primary, secondary and laccase-treated secondary emulsion (**Figure 6A**). The ζ-potential of the fish gelatin-stabilized droplets in the primary emulsion remained positive (+18.9 ± 1.2 mV) regardless of freeze-thaw cycles, and the ζ-potential of the fish gelatin-beet pectin coated droplets in the secondary and enzyme-treated emulsions remained negative (-21.5 ± 0.7 mV). There was a slight increase in the mean particle size of primary emulsion after two freeze-thaw cycles ($p < 0.05$) (**Figure 6B**). The mean particle diameter increased from 222 ± 6 nm at zero cycles to 300 ± 26 nm after two cycles. Although the mean particle diameter remained almost constant, coalesced oil was visible in fish gelatin-stabilized oil droplets of primary emulsions (**Figure 7A** and **7B**). Layers formed by random-coil proteins are known to be less resistant to rupture than those formed by globular proteins such as whey proteins (*McClements*, 2004; *McClements et al.*, 2009). Ice crystals which are formed during freezing may physically penetrate the interfacial membrane making oil droplets more prone to coalescence when the emulsion is again thawed (*Ogawa et al.*, 2003; *Guzey et al.*, 2006), a fact that has been previously described by *Klinkesorn et al.* (2005) and *Aoki et al.* (2005).

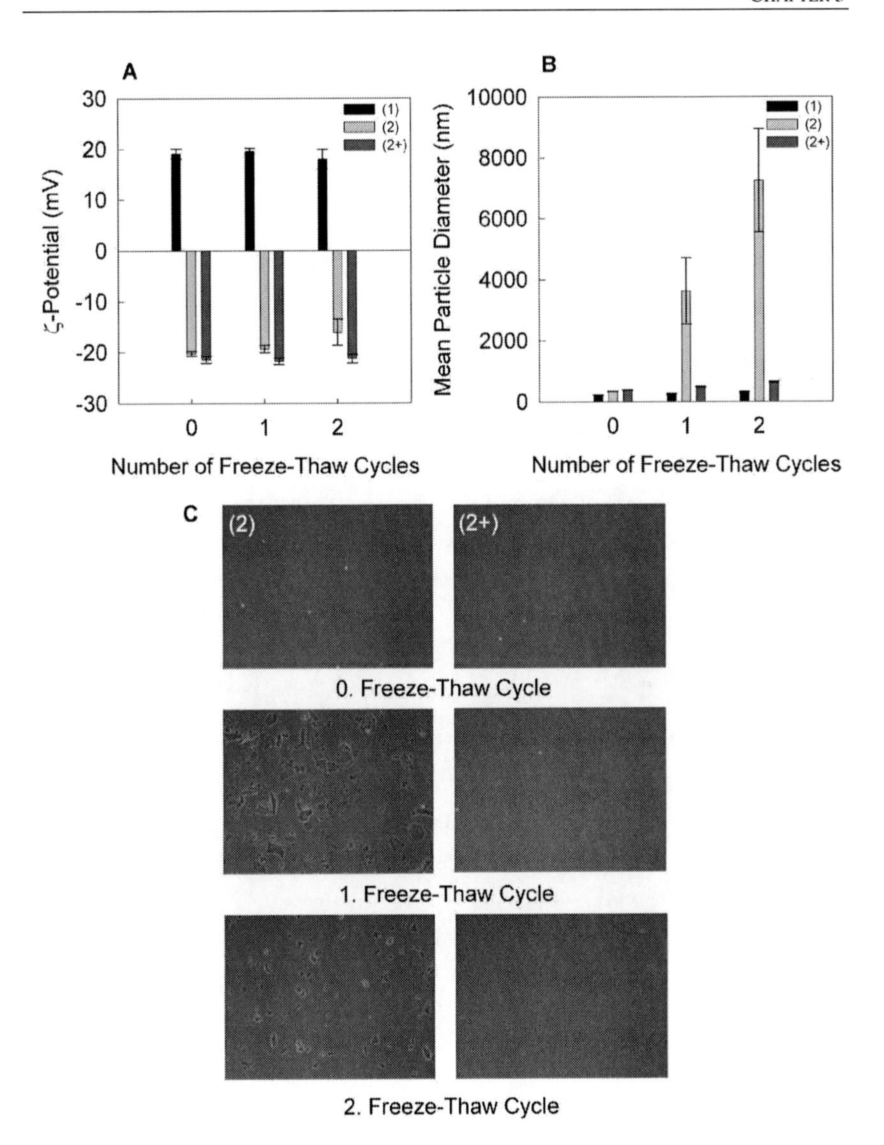


**Figure 6 Dependence of the electrical charge (ζ-potential, A), mean particle diameter (B), and microstructure (C) on number of freeze-thaw cycles of fish gelatin-coated (1), beet pectin-fish gelatin-coated (2), and laccase crosslinked beet pectin-fish gelatin-coated emulsion droplets (2+).**

SDS-coated oil droplets displayed extensive oiling-off after a single freeze-thaw cycle. The fish gelatin-sugar beet pectin-stabilized emulsion exhibited extensive droplet aggregation after the first cycle, which led to rapid droplet creaming (**Figure 6C**). Similar results have previously been published for other secondary emulsions (*Aoki et al.*, 2005; *Klinkesorn et al.*,

2005). In contrast, the mean particle diameter of laccase-treated secondary emulsion increased only little (**Figure 6B** and **6C**) and freeze-thaw studies with oil-in-water emulsions stabilized by cross-linked beet pectin also showed no coalescence and oiling-off (**Figure 7C** and **7D**). The improved stability of laccase-treated emulsion maybe due to the formation of diferulic covalent bonds between the beet pectin molecules which make the interfacial membrane more resistant to rupture by oil and ice crystals or mechanical forces that tend to push the droplets together during the freezing.

**Key insights**

In summary, we can highlight a number of key insights from our study:

- Fish gelatin may be used as an emulsifier to form surprisingly stable oil-in-water emulsions with a single membrane layer, because fish gelatin acts as a steric barrier to prevent droplet aggregation or creaming caused by high salt levels or temperatures. However, fish gelatin coated oil droplets are prone to coalescence after freezing and thawing.

- A double-layer interfacial membrane consisting of fish gelatin – beet pectin may improve creaming stability of emulsions. Additionally, these multilayered emulsions show enhanced freeze-thaw stabilities.

- Laccase promotes an enzymatic crosslinking of adsorbed sugar beet pectin in multilayered emulsions thereby improving stability of emulsions to heating. Moreover, laccase-treated emulsions withstand freeze-thaw cycles without any sign of (partial) coalescence and/or aggregation.

- Creaming experiments under normal gravitational conditions at room temperature over a period of 10 days showed no evidence of creaming in any of the manufactured emulsions which is likely due to the small size of the oil droplets in primary, secondary, and enzyme-treated secondary emulsions (*Wooster et al.*, 2008).

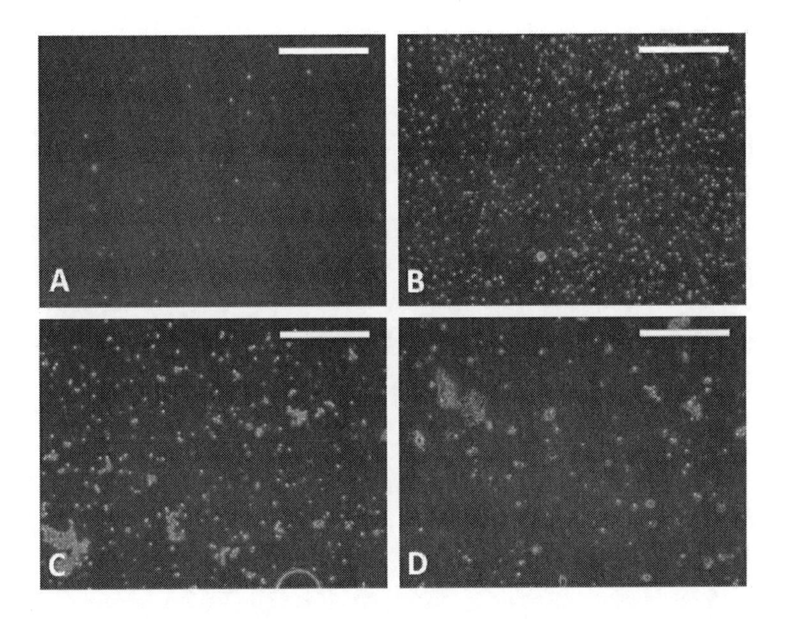


**Figure 7 Dependence of coalescences on freeze-thaw temperature of fish gelatin-stabilized oil droplets (before (A) and after (B) cycle) and beet pectin-stabilized oil droplets (before (C) and after (D) cycle). Fish gelatin-coated emulsions were incubated in a -8 °C water bath for 22 h and thawed by incubating them in a 25 °C water bath for 2 h (scale bar 50 µm). Beet pectin-coated emulsions were incubated in a -22 °C refrigerator for 22 h and thawed by incubating them in a 25 °C water bath for 2 h. Coalesced oil droplets appear as bright "dots".**

## CONCLUSIONS

The study demonstrated that enzymatic crosslinking is a valuable tool to further enhance the functional properties of multilayered emulsions. A number of enzymes in addition to laccase could be used to achieve such an effect, e.g. transglutaminase. In particular, in multilayered systems composed of both proteins and carbohydrates, a combination of a pectin crosslinker and a carbohydrate crosslinker could lead to the formation of extremely elastic and resistant membranes, a hypothesis that should be tested in future studies. Our results have important implications for food manufacturer that are in need of having emulsions with improved resistance, e.g. emulsions used in frozen foods for sauces, dips, etc. The above described approach could enable them to create foods that maintain their quality despite destabilizing storage and transport conditions.

## ACKNOWLEDGEMENTS

We would like to thank Herbstreith & Fox KG (Neuenbürg, Germany) for generously providing us with pectin samples. We appreciate the help and fruitful discussions with Dr. Monika Gibis (University of Hohenheim).

## REFERENCES


Aoki, T., Decker, E. A., et al. (2005): Influence of environmental stresses on stability of O/W emulsions containing droplets stabilized by multilayered membranes produced by a layer-by-layer electrostatic deposition technique. Food Hydrocolloids 19(2): 209-220.

Ausborn, M., Schreier, H., et al. (1994): The protective effect of free and membrane-bound cryoprotectants during freezing and freeze-drying of liposomes. Journal of Controlled Release 30(2): 105-116.

Avena-Bustillos, R. J., Olsen, C. W., et al. (2006): Water vapor permeability of mammalian and fish gelatin films. Journal of Food Science 71(4): E202-E207.

Babel, W. (1996): Gelatine – ein vielseitiges Biopolymer. Chemie in unserer Zeit 30(2): 86-95.

Belitz, H.-D., Grosch, W., et al. (2001): Lehrbuch der Lebensmittelchemie. Berlin, Springer-Verlag GmbH.

Burjanadze, T. V. (2000): New analysis of the phylogenetic change of collagen thermostability. Biopolymers 53(6): 523-528.

Chanamai, R., McClements, D. J. (2001): Depletion flocculation of beverage emulsions by gum arabic and modified starch. Journal of Food Science 66(3): 457-463.

Chanamai, R., McClements, D. J. (2002): Comparison of gum arabic, modified starch, and whey protein isolate as emulsifiers: Influence of pH, $CaCl_2$ and temperature. Journal of Food Science 67(1): 120-125.

Chen, B., McClements, D. J., et al. (2010): Stabilization of soybean oil bodies by enzyme (Laccase) cross-linking of adsorbed beet pectin coatings. Journal of Agricultural and Food Chemistry 58(16): 9259-9265.

Claus, H. (2004): Laccases: Structure, reactions, distribution. Micron 35(1-2): 93-96.

Cornacchia, L., Roos, Y. H. (2011): Lipid and water crystallization in protein-stabilized oil-in-water emulsions. Food Hydrocolloids 25(7): 1726-1736.

Dalgleish, D. G., Hallett, F. R. (1995): Dynamic light scattering: Applications to food systems. Food Research International 28(3): 181-193.

Decher, G., Hong, J. D., et al. (1992): Buildup of ultrathin multilayer films by a self-assembly process: III. Consecutively alternating adsorption of anionic and cationic polyelectrolytes on charged surfaces. Thin Solid Films 210-211(PART 2): 831-835.

Demetriades, K., Coupland, J. N., et al. (1997): Physical properties of whey protein stabilized emulsions as related to pH and NaCl. Journal of Food Science 62(2): 342-347.

Demetriades, K., Coupland, J. N., et al. (1997): Physicochemical properties of whey protein-stabilized emulsions as affected by heating and ionic strength. Journal of Food Science 62(3): 462-467.

Dickinson, E. (1992): Faraday research article. Structure and composition of adsorbed protein layers and the relationship to emulsion stability. Journal of the Chemical Society, Faraday Transactions 88(20): 2973-2983.

Dickinson, E. (1994): Protein-stabilized emulsions. Journal of Food Engineering 22(1-4): 59-74.

Dickinson, E. (2003): Hydrocolloids at interfaces and the influence on the properties of dispersed systems. Food Hydrocolloids 17(1): 25-39.

Dickinson, E. (2010): Flocculation of protein-stabilized oil-in-water emulsions. Colloids and Surfaces B: Biointerfaces.

Funami, T., Zhang, G., et al. (2007): Effects of the proteinaceous moiety on the emulsifying properties of sugar beet pectin. Food Hydrocolloids 21(8): 1319-1329.

Grigoriev, D. O., Miller, R. (2009): Mono- and multilayer covered drops as carriers. Current Opinion in Colloid & Interface Science 14(1): 48-59.

Gu, Y. S., Decker, A. E., et al. (2005): Production and characterization of oil-in-water emulsions containing droplets stabilized by multilayer membranes consisting of β-lactoglobulin, ι-carrageenan and gelatin. Langmuir 21(13): 5752-5760.

Gu, Y. S., Decker, E. A., et al. (2004): Influence of pH and ι-carrageenan concentration on physicochemical properties and stability of β-lactoglobulin-stabilized oil-in-water emulsions. Journal of Agricultural and Food Chemistry 52(11): 3626-3632.

Guzey, D., Kim, H. J., et al. (2004): Factors influencing the production of o/w emulsions stabilized by β-lactoglobulin-pectin membranes. Food Hydrocolloids 18(6): 967-975.

Guzey, D., McClements, D. (2006): Influence of environmental stresses on o/w emulsions stabilized by β-lactoglobulin–pectin and β-lactoglobulin–pectin–chitosan membranes produced by the electrostatic layer-by-layer deposition technique. Food Biophysics 1(1): 30-40.

Guzey, D., McClements, D. J. (2006): Formation, stability and properties of multilayer emulsions for application in the food industry. Advances in Colloid and Interface Science 128-130: 227-248.

Harada, T., Yokomizo, K. (2000): Demulsification of oil-in-water emulsion under freezing conditions: Effect of crystal structure modifier. JAOCS, Journal of the American Oil Chemists' Society 77(8): 859-863.

Harrington, J. C., Morris, E. R. (2009): Conformational ordering and gelation of gelatin in mixtures with soluble polysaccharides. Food Hydrocolloids 23(2): 327-336.

Iwanaga, D., Gray, D., et al. (2008): Stabilization of soybean oil bodies using protective pectin coatings formed by electrostatic deposition. Journal of Agricultural and Food Chemistry 56(6): 2240-2245.

Klinkesorn, U., Sophanodora, P., et al. (2005): Encapsulation of emulsified tuna oil in two-layered interfacial membranes prepared using electrostatic layer-by-layer deposition. Food Hydrocolloids 19(6): 1044-1053.

Komatsu, H., Okada, S., et al. (1997): Suppressive effects of salts on droplet coalescence in a commercially available fat emulsion during freezing for storage. Journal of Pharmaceutical Sciences 86(4): 497-502.

Leroux, J., Langendorff, V., et al. (2003): Emulsion stabilizing properties of pectin. Food Hydrocolloids 17(4): 455-462.

Littoz, F., McClements, D. J. (2008): Bio-mimetic approach to improving emulsion stability: Cross-linking adsorbed beet pectin layers using laccase. Food Hydrocolloids 22(7): 1203-1211.

McClements, D. J. (2004): Food emulsions: Principles, practice, and techniques. Boca Raton, CRC Press.

McClements, D. J. (2004): Protein-stabilized emulsions. Current Opinion in Colloid & Interface Science 9(5): 305-313.

McClements, D. J., Decker, E. A., et al. (2009): Structural design principles for delivery of bioactive components in nutraceuticals and functional foods. Critical Reviews in Food Science and Nutrition 49(6): 577 - 606.

McClements, D. J., Decker, E. A., et al. (2007): Emulsion-based delivery systems for lipophilic bioactive components. Journal of Food Science 72(8): R109-R124.

McClements, D. J., Rao, J. (2011): Food-grade nanoemulsions: Formulation, fabrication, properties, performance, biological fate, and potential toxicity. Critical Reviews in Food Science and Nutrition 51(4): 285-330.

McClements, D. J., Stefan, K., et al. (2009). Biopolymers in food emulsions. Modern Biopolymer Science. San Diego, Academic Press: 129-166.

Minussi, R. C., Pastore, G. M., et al. (2002): Potential applications of laccase in the food industry. Trends in Food Science & Technology 13(6-7): 205-216.

Moreau, L., Kim, H. J., et al. (2003): Production and characterization of oil-in-water emulsions containing droplets stabilized by □-lactoglobulin-pectin membranes. Journal of Agricultural and Food Chemistry 51(22): 6612-6617.

Morris, G. A., Ralet, M.-C., et al. (2010): Physical characterization of the rhamnogalacturonan and homogalacturonan fractions of sugar beet (Beta vulgaris) pectin. Carbohydrate Polymers 82(4): 1161-1167.

Ogawa, S., Decker, E. A., et al. (2003): Influence of environmental conditions on the stability of oil in water emulsions containing droplets stabilized by lecithin-chitosan membranes. Journal of Agricultural and Food Chemistry 51(18): 5522-5527.

Onsaard, E., Vittayanont, M., et al. (2006): Comparison of properties of oil-in-water emulsions stabilized by coconut cream proteins with those stabilized by whey protein isolate. Food Research International 39(1): 78-86.

Rombouts, F. M., Thibault, J. F. (1986): Feruloylated pectic substances from sugar-beet pulp. Carbohydrate Research 154(1): 177-187.

Saito, H., Kawagishi, A., et al. (1999): Coalescence of lipid emulsions in floating and freeze-thawing processes: Examination of the coalescence transition state theory. Journal of Colloid and Interface Science 219(1): 129-134.

Strauss, G., Hauser, H. (1986): Stabilization of lipid bilayer vesicles by sucrose during freezing. Proceedings of the National Academy of Sciences of the United States of America 83(8): 2422-2426.

Surh, J. (2007): Influence of pH, emulsifier concentration, and homogenization condition on the production of stable oil-in-water emulsion droplets coated with fish gelatin. Food Science and Biotechnology 16(6): 999-1005.

Surh, J., Decker, E. A., et al. (2006): Properties and stability of oil-in-water emulsions stabilized by fish gelatin. Food Hydrocolloids 20(5): 596-606.

Thurston, C. F. (1994): The structure and function of fungal laccases. Microbiology 140(1): 19-26.

Voragen, F., Beldman, G., et al. (2008). Chemistry and enzymology of pectins. Advanced Dietary Fibre Technology. L. P. Barry V. McCleary: 379-398.

Williams, P. A., Sayers, C., et al. (2005): Elucidation of the emulsification properties of sugar beet pectin. Journal of Agricultural and Food Chemistry 53(9): 3592-3597.

Wooster, T. J., Golding, M., et al. (2008): Impact of oil type on nanoemulsion formation and Ostwald ripening stability. Langmuir 24(22): 12758-12765.

Zeeb, B., Gibis, M., et al. (2012): Crosslinking of interfacial layers in multilayered oil-in-water emulsions using laccase: Characterization and pH-stability. Food Hydrocolloids 27(1): 126-136.

# CHAPTER 4

## Influence of interfacial properties on Ostwald ripening in crosslinked multilayered oil-in-water emulsions

*Benjamin Zeeb[1], Monika Gibis[1], Lutz Fischer[2], Jochen Weiss[1]*

[1] Department of Food Physics and Meat Science, University of Hohenheim, Garbenstrasse 21/25, 70599 Stuttgart, Germany

[2] Department of Food Biotechnology, University of Hohenheim, Garbenstrasse 25,70599 Stuttgart, Germany

Reprinted from "*Influence of interfacial properties on Ostwald ripening in crosslinked multilayered oil-in-water emulsions*", Zeeb, B., Gibis, M., Fischer, L., Weiss, J., Journal of Colloid and Interface Science, 2012, 387(1), p. 65-73 with permission from Elsevier.

## ABSTRACT

The influence of interfacial crosslinking, layer thickness and layer density on the kinetics of Ostwald ripening in multilayered emulsions at different temperatures was investigated. Growth rates of droplets were measured by monitoring changes in the droplet size distributions of 0.5% (w/w) *n*-octane, *n*-decane, and *n*-dodecane oil-in-water emulsions using static light scattering. Lifshitz-Slyozov-Wagner theory was used to calculate Ostwald ripening rates. A sequential two step process, based on electrostatic deposition of sugar beet pectin onto fish gelatin or whey protein isolate (WPI) interfacial membranes, was used to manipulate the interfacial properties of the oil droplets. Laccase was added to the fish gelatin-beet pectin emulsions to promote crosslinking of adsorbed pectin molecules via ferulic acid groups, whereas heat was induced to promote crosslinking of WPI and helix coil transitions of fish gelatin. Ripening rates of single-layered, double-layered and crosslinked emulsions increased as the chain length of the *n*-alkanes decreased. Emulsions containing crosslinked fish gelatin-beet pectin coated droplets had lower droplet growth rates ($3.1 \pm 0.3 \times 10^{-26}$ $m^3/s$) than fish gelatin-stabilized droplets ($7.3 \pm 0.2 \times 10^{-26}$ $m^3/s$), which was attributed to the formation of a protective network. Results suggest that physical or enzymatic biopolymer-crosslinking of interfaces may reduce the molecular transport of alkanes between the droplets in the continuous phase.

**Keywords:** Multilayered emulsions; Layer-by-layer electrostatic deposition method; Laccase; Ostwald ripening; Layer thickness; Layer density

## INTRODUCTION

Some foods, cosmetics, pharmaceuticals, or paints are natural or manufactured emulsion based products which may undergo a number of different processes leading to physical instability such as creaming, sedimentation, flocculation, coalescence, partial coalescence, phase inversion, and Ostwald ripening (*McClements*, 2004; *McClements et al.*, 2009). In particular, coalescence and Ostwald ripening may play a substantial role in the breakdown of emulsions due to an irreversible decrease in their dispersity (*Kabalnov et al.*, 1992). Coalescence is the process by which two or more droplets fuse together to form one single larger droplet (*Weers*, 1998). In contrast, Ostwald ripening also referred to as molecular diffusion or isothermal distillation is the process whereby larger droplets grow at the expense of smaller ones (*Voorhees*, 1985; *Weiss et al.*, 1999; *McClements*, 2004). The driving force for the mass transport is the difference in the chemical potential or dispersed phase solubility between differently sized droplets which is described by the Kelvin Equation (*Weers*, 1998):

$$c(r) = c(\infty)exp\frac{2\gamma^i V_m}{rRT} \tag{1}$$

where $c(r)$ is the solubility of a compound in the immediate vicinity of a particle of radius $r$, $c(\infty)$ is the bulk phase solubility, $\gamma^i$ is the interfacial tension between the dispersed and continuous phases, $V_m$ is the molar volume of the dispersed phase, $R$ is the molar gas constant, and $T$ is the absolute temperature. The kinetics of Ostwald ripening are well described in terms of the theory which was first developed by Lifshitz and Slyozov and independently by Wagner (LSW theory) (*Lifshitz et al.*, 1961; *Kabalnov et al.*, 1987; *Kabalnov et al.*, 1992; *Weers*, 1998; *Weiss et al.*, 1999; *Solans et al.*, 2005). A major concern of manufacturers and scientists is therefore to better understand the factors that may influence Ostwald ripening so as to slow its rate or to prevent it from occurring in order to prolong the shelf life of the final product (*Taylor*, 1998; *Weers*, 1998; *Kabalnov*, 2001).

Various factors have been studied that may retard Ostwald ripening in emulsions such as droplet size distribution, solute solubility, interfacial tension, and droplet composition (*De Smet et al.*, 2000; *McClements*, 2004). It has been described that the destabilization of emulsions by Ostwald ripening depends on the chain length of the dispersed alkane phase. *Weiss et al.* (1999) and *Dickinson et al.* (1999) found an exponential increase of the growth rate as the *n*-alkane chain length or the molar volume of the alkane decreased linearly.

Furthermore, *Davis et al.* (1981) showed that emulsions prepared by homogenizing short-chain alkanes could be stabilized with respect to Ostwald ripening by the addition of small amounts of long-chain alkanes. Several studies focused on the role of excess surfactant or alcohols in the continuous phase, facilitating the mass transfer of alkanes between the droplets. For example, it has been demonstrated that Ostwald ripening rates increased with increasing surfactant concentrations since the number of micelles acting as carriers for the oil molecules increases (*Soma et al.*, 1996; *Weiss et al.*, 2000; *Solans et al.*, 2005). *Weiss et al.* (1999) also indicated that the rate of molecular diffusion was affected by the molecular structure of the surfactants. In case of nanoemulsions, it was described that the physical properties of the oil phase was found to have a considerable impact on the emulsions formation and stabilization (*Wooster et al.*, 2008). Ethanol has been known to decrease the surface tension, therefore facilitating Ostwald ripening of protein-stabilized emulsions (*Dickinson et al.*, 1998; *Dickinson et al.*, 1999). Furthermore, emulsions having high oil droplet concentrations typically have broader particle size distributions and therefore higher growth rates than those predicted by LSW theory due to an increased mass transport between particles which are in close proximity (*Weers*, 1998). It was also proposed that Ostwald ripening may be reduced by adding thickening or gelling agents to the continuous phase to increase its viscosity or to form a gel. However, the agent used to modify the continuous phase has to introduce a yield stress which is larger than the Laplace pressure of the oil droplets, an approach that is not considered to be practical (*Weers*, 1998).

Some studies focused on the formation of thick and resistant interfacial membranes surrounding the droplets in order to retard Ostwald ripening (*Kabalnov et al.*, 1992; *De Smet et al.*, 2000; *Meinders et al.*, 2001; *McClements*, 2004; *Meinders et al.*, 2004). *McClements* (2004) proposed that the growth rate of droplets increases as the interfacial tension increases (see Equation (1)). If the interfacial membrane resists the shrinkage and expansion that have to occur for the ripening process to take place, the overall growth of the droplets may be reduced. To this purpose, some theoretical analysis of the influence of interfacial rheology in Ostwald ripening has been carried out by *Meinders et al.* (2001, 2004) that indicated that higher elasticities in membranes may reduce Ostwald ripening. Along the same line, it was hypothesized that Ostwald ripening could be slowed down by the use of emulsifiers that are (i) strongly adsorbed to the interface and (ii) do not readily desorb during the ripening process (*Weers*, 1998; *Wooster et al.*, 2008).

A well-established interfacial engineering technology, based on the electrostatic deposition of charged biopolymers onto surfaces of oppositely charged particles was developed to create emulsions with improved resistance to environmental stresses (*Gu et al.*, 2004; *Ogawa et al.*, 2004; *Gu et al.*, 2005; *Guzey et al.*, 2006). These so called "multilayered" emulsion have similar bulk physicochemical properties such as viscosity and appearance as conventional emulsions with similar particle characteristics (e.g. concentration, size distribution, and net charge) (*Guzey et al.*, 2006; *McClements et al.*, 2010). However, layer-by-layer electrostatically coated droplets with multilamellar shells did not destabilize during freeze-thaw cycling or dehydration (*Ogawa et al.*, 2003; *Gu et al.*, 2004; *Aoki et al.*, 2005; *Guzey et al.*, 2006; *Littoz et al.*, 2008). *Mun et al.* (2006) already demonstrated that the stability of emulsions to Ostwald ripening could be improved by using the electrostatic deposition method. In their study, SDS-chitosan coated *n*-tetradecane droplets grew little, whereas single SDS-stabilized droplets showed a remarkable increase in mean droplet size with time.

We recently carried out a series of studies based on the work of *Littoz et al.* (2008) to further improve the functionality of multilayered emulsions by enzymatically crosslinking the interfacial membranes (*Littoz et al.*, 2008; *Zeeb et al.*, 2011; *Zeeb et al.*, 2012). Generally, physical, chemical, or enzymatic approaches may be used to covalently crosslink biopolymers in interfacial membranes thereby increasing the resistance of emulsions to break down over time (*Romoscanu et al.*, 2005; *Romoscanu et al.*, 2006; *McClements et al.*, 2009). Several biomimetic approaches have been carried out to investigate the use of enzymes to crosslink biopolymers, including transglutaminase or laccase (*Dickinson*, 1997; *Littoz et al.*, 2008; *Zeeb et al.*, 2011; *Zeeb et al.*, 2012). An early study by *Dickinson et al.* (1999) indicated that transglutaminase-induced interfacial crosslinking of adsorbed protein films surrounding droplets in oil-in-water emulsions may be able to provide some limited improvement in emulsion stability to Ostwald ripening. In our studies, we specifically showed that emulsions containing a layer of electrostatically deposited pectin had significantly improved abilities to resist addition of salts or changes in pH after treatment with laccase (*Zeeb et al.*, 2011; *Zeeb et al.*, 2012). Overall, application of enzymes is considered to be a mild processing technology and could be implemented using conventional processing technologies such as mixing or homogenization (*Minussi et al.*, 2002; *Zeeb et al.*, 2011).

The objective of the present study was to establish a better understanding of the effect of crosslinking on the process of Ostwald ripening in emulsions. To this purpose, we studied the change in size of primary, secondary (coated), and laccase-treated secondary emulsions induced by molecular diffusion using alkanes with different solubilities and at different temperatures. We hypothesized that the thicker and denser the interfacial membrane is, the less susceptible to Ostwald ripening the emulsion may be, since there should be an increased resistance to pressure changes in the droplets induced by the mass transport between them. *N*-alkane-in-water emulsions were fabricated with a single- or double-layered membrane, treated with laccase, and subjected to various storage temperatures. To study the influence of membrane density on Ostwald ripening, we used a highly flexible or a globular protein as emulsifier, namely fish gelatin and whey protein isolate (WPI). Both of these primary emulsifiers could be coated with negatively charged sugar beet pectin at low pH values.

## MATERIALS AND METHODS

**Materials.** Cold water fish skin gelatin (#049K0050) was purchased from Sigma-Aldrich Co. (Steinheim, Germany). Its average molecular weight and p*I* value were reported to be approximately 60 kDa and pH 6, respectively. Whey protein isolate (#B180214) was donated by Arla Foods Ingredients (Viby, Denmark). Sugar beet pectin (#1 09 03 135) was a gift of Herbstreith & Fox KG (Neuenbürg, Germany). As stated by the manufacturer the degree of esterification of the beet pectin was 55%. All biopolymers were used without further purification. Octane (#07696APV, purity $\geq$ 99.0%), decane (#08096LMV, purity $\geq$ 99.0%), and dodecane (#65796EM, purity $\geq$ 99.0%) were obtained from Sigma-Aldrich Co. (Steinheim, Germany). The enzyme laccase (#0001437590, from *Trametes versicolor*) was also purchased from Sigma-Aldrich Co. (Steinheim, Germany). Laccase was reported to have 20.7 units per mg (AU) of enzyme. Citric acid monohydrate (#409107294) was obtained from Carl Roth GmbH & Co. KG (Karlsruhe, Germany) and sodium citrate dihydrate (#26996TH) was purchased from SAFC (St. Louis, MO). Analytical grade hydrochloric acid (HCl) and sodium hydroxide (NaOH) were purchased from Carl Roth GmbH & Co. KG (Karlsruhe, Germany). Distilled water was used for the preparation of all samples.

**Solution preparation.** An aqueous emulsifier solution was prepared by dispersing 1% (w/w) fish gelatin powder or 1% (w/w) whey protein isolate powder, respectively, into 10 mM citrate buffer (pH 3.5). Sugar beet pectin solution was prepared by dispersing 2% (w/w) powdered pectin into 10 mM citrate buffer at pH 3.5. All solutions were stirred at least two hours to ensure complete hydration and then adjusted to a pH of 3.5 using 1 M HCl and/or 1

M NaOH. Enzyme solution was prepared by dispersing enzyme powder into 10 mM citrate buffer (pH 3.5).

**Primary emulsion preparation.** Primary emulsions were prepared by homogenizing 5% (w/w) *n*-alkane (octane, decane, dodecane) with 95% (w/w) aqueous fish gelatin or whey protein isolate solution, respectively, using a high shear blender (Standard Unit, IKA Werk GmbH, Germany) for 2 min followed by three passes at 10000 psi (68.95 MPa) through a high pressure homogenizer (M110-EH-30, Microfluidics International Cooperation, Newton, MA). Primary emulsions were diluted with citrate buffer (10 mM, pH 3.5) to obtain a final oil droplet concentration of 0.5% (w/w).

**Secondary emulsion preparation.** Stock primary emulsions were diluted (1:10) with 10 mM citrate buffer (pH 3.5) and aqueous beet pectin solution to form secondary emulsions (0.2% (w/w) beet pectin) with an oil droplet content of 0.5% (w/w).

**Enzyme-induced crosslinking of secondary emulsions.** Laccase-treated secondary emulsions were prepared by mixing secondary emulsions with citrate buffer (pH 3.5) and the enzyme solution using a vortex. An enzyme/beet pectin ratio of 0.24 mg/4 mg, equivalent to 5 AU, was sufficient to promote interfacial crosslinking of sugar beet pectin as previously described by *Zeeb et al.* (2012).

**Heat-induced modifications of secondary emulsion.** In addition to enzymatic biopolymer crosslinking of interfaces, a physically induced transformation was also carried out. Dodecane was selected as a dispersed phase fraction due to its low water solubility in comparison to octane or decane. Secondary dodecane-in-water emulsions were prepared as described above. To promote crosslinking of adsorbed WPI or fish gelatin layers, secondary emulsions were subjected to isothermal heat treatment at 80 °C for 10 min according to *Romoscanu et al.* (2006) and *McClements et al.* (1993).

**Determination of droplet size changes.** The particle size distributions of emulsions were measured using a static light scattering instrument (Horiba LA-950, Retsch Technology GmbH, Haan, Germany). Samples were withdrawn and diluted to a droplet concentration of approximately 0.005% (w/w) with an appropriate buffer to prevent multiple scattering effects. The instrument measured the angular dependence of the intensity of the laser beam scattered by the dilute emulsions and then used the Mie theory to calculate the droplet size distributions that gave the best fit between theoretical predictions and empirical measurements. A refractive index ratio of 1.06 (ratio of the indices between the oil and water phase) was used. The particle size measurements are reported as the number-average droplet diameter $d_{10}$ which is defined by $\frac{1}{n}\sum_i n_i d$, where $n_i$ is the number of droplets of diameter $d_i$ and $n$ is the total number of droplets. Particle radii were calculated as the average of measurements made on at least two freshly prepared samples. All emulsions were stored at either 10 °C, 22 °C, or 55 °C using a temperature controlled cabinet and the droplet size distribution was measured at regular time intervals.

**Statistical analysis.** All experiments were repeated at least two times using freshly prepared samples. Means and standard deviations were calculated from a minimum of three measurements using Excel. Results were analyzed with a statistical software (SAS 9.2, SAS Institute, Cary, NC). Linear regression analysis was used to determine slope, standard error, and coefficient of determination.

## RESULTS AND DISCUSSION

### Theoretical considerations – The LSW theory

The LSW theory is valid under the following assumptions: (i) the particles dispersed in the continuous phase are spherical and fixed in space; (ii) the particles are separated from each other by distances where no interactions between neighbouring particles will take place; (iii) molecular diffusion limits the mass transport in the dispersion medium; (iv) the concentration of the molecularly dissolved material is constant except for the domains surrounding the particles with a thickness of about $r$ (*Kabalnov et al.*, 1992; *Soma et al.*, 1996; *Weers*, 1998). Low oil droplet concentrations of 0.5% (w/w) were therefore selected in order to meet these conditions. With these assumptions, the Ostwald ripening rate $\omega$ has been derived by Lifshitz and Slyozov and independently by Wagner as a function of a number of physical parameters (*Weers*, 1998; *Weiss et al.*, 1999; *Sakai et al.*, 2002):

$$\omega = \frac{dr_c^3}{dt} = \frac{8\gamma^i D_0 c(\infty) V_m^2}{9RT} \tag{2}$$

where $r_c$ is the critical radius of a droplet, and $D_0$ is the diffusion coefficient for the dispersed phase molecules in the continuous phase. It follows then that droplets with $r > r_c$ grow at the expense of smaller ones, whereby droplets with $r < r_c$ tend to disappear (*Weers*, 1998). Equation (2) indicates that the droplet volume (proportional to $r_c^3$) increases linearly with time. Diffusion coefficients for spherical particles are temperature dependent, a fact that has been described by the Stokes-Einstein equation (*Hiemenz et al.*, 1997 ):

$$D_0 = \frac{k_B T}{6\pi\eta r} \tag{3}$$

where $D_O$ is the bulk diffusion coefficient at temperature $T$ for the dispersed phase, $k_B$ is the Boltzmann Constant, $\eta$ is the dynamic viscosity and $r$ is the droplet radius. In order to calculate the bulk diffusion coefficients for any of the prepared *n*-alkane oil-in-water emulsions stabilized by different interfacial membranes and stored at different temperatures, the values need to be adjusted. By Equation (3), the bulk diffusion coefficient $D_O$ is proportional to $T/\eta$ which could be applied to yield the following relationship (*Hiemenz et al.*, 1997 ):

$$\frac{D_{OS}}{D_{OT}} = \frac{T_S/\eta_S}{T_{ex}/\eta_{ex}} = \frac{T_S}{T_{ex}}\frac{\eta_{ex}}{\eta_S} \tag{4}$$

where $D_{OT}$ is the bulk diffusion coefficient of a *n*-alkane measured in an experiment (subscript *ex*) at a given temperature $T_{ex}$ at which the dynamic viscosity of the solvent is $\eta_{ex}$. It should be noted that the simplified models described above may not be entirely accurate as multilayered emulsion droplets may not be completely spherical. Nevertheless, the application of the models does allow a comparison of the obtained data. The physical characteristics of *n*-octane, *n*-decane, and *n*-dodecane used in our experiments and their bulk diffusion coefficients are shown in **Table 1**. Dynamic viscosity values $\eta_s$ at standard conditions (T = 20 °C; subscript *s*) of the aqueous phase are shown in **Table 2** and were taken elsewhere (*Mamedov et al.*, 1973). Based on Equations (2) and (4) and the Ostwald ripening rates $\omega$ determined from linear regression (**Table 3**), the diffusion coefficients of primary, secondary, and enzyme-treated secondary *n*-alkane-in-water emulsions were recalculated and are shown in **Tables 4** and **5**.

**Table 1 Physical characteristics of *n*-alkanes used in the experiments*.**

| property | | unit | octane | decane | dodecane |
|---|---|---|---|---|---|
| interfacial tension | $\gamma$ | N/m | $5.168 \times 10^{-2}$ | $5.230 \times 10^{-2}$ | $5.278 \times 10^{-2}$ |
| water solubility | $c(\infty)$ | $mol/m^3$ | $5.8 \times 10^{-3}$ | $3.6 \times 10^{-4}$ | $2.3 \times 10^{-5}$ |
| molar volume | $V_m$ | $m^3/mol$ | $1.63 \times 10^{-4}$ | $1.94 \times 10^{-4}$ | $2.27 \times 10^{-4}$ |
| molecular weight | $M_w$ | kg/mol | 0.114 | 0.142 | 0.170 |
| molar gas constant | R | J/mol*K | 8.314 | 8.314 | 8.314 |

* Data obtained from *Weiss et al.* (1999) and *Sakai et al.* (2002).

**Table 2 Adjusted bulk diffusion coefficients of *n*-octane, *n*-decane, and *n*-dodecane according to equation (4) (standard conditions T = 20 °C).**

| temperature (°C) | temperature (K) | dynamic viscosity of aqueous phase* $\eta$ ($N*s/m^2$) | bulk diffusion coefficient of octane $D_0$ ($m^2/s$) | bulk diffusion coefficient of decane $D_0$ ($m^2/s$) | bulk diffusion coefficient of dodecane $D_0$ ($m^2/s$) |
|---|---|---|---|---|---|
| 10 | 283.15 | $1.308 \times 10^{-3}$ | $5.8 \times 10^{-10}$ | $4.4 \times 10^{-10}$ | $4.0 \times 10^{-10}$ |
| 20 | 293.15 | $1.003 \times 10^{-3}$ | $7.8 \times 10^{-10}$ | $5.9 \times 10^{-10}$ | $5.4 \times 10^{-10}$ |
| 22 | 295.15 | $0.955 \times 10^{-3}$ | $8.2 \times 10^{-10}$ | $6.2 \times 10^{-10}$ | $5.7 \times 10^{-10}$ |
| 55 | 328.15 | $0.504 \times 10^{-3}$ | $9.8 \times 10^{-10}$ | $7.4 \times 10^{-10}$ | $6.8 \times 10^{-10}$ |

*Dynamic viscosity values were taken from *Mamedov et al.* (1973).

**Table 3 Measured Ostwald ripening rates depending on *n*-alkane type, temperature, and interface treatment using fish gelatin as primary emulsifier.**

| *n*-alkane | temperature (°C) | primary o/w OR rate ($m^3/s$) | secondary o/w OR rate ($m^3/s$) | crosslinked o/w OR rate ($m^3/s$) |
|---|---|---|---|---|
| octane | 10 | $6.0 \pm 0.4 \times 10^{-24}$ | $7.6 \pm 0.2 \times 10^{-24}$ | $4.7 \pm 0.2 \times 10^{-24}$ |
| | 22 | $8.9 \pm 0.2 \times 10^{-24}$ | $1.2 \pm 0.0 \times 10^{-23}$ | $6.7 \pm 0.2 \times 10^{-24}$ |
| | 55 | $1.6 \pm 0.1 \times 10^{-23}$ | $2.0 \pm 0.1 \times 10^{-23}$ | $1.8 \pm 0.0 \times 10^{-23}$ |
| decane | 10 | $4.3 \pm 0.1 \times 10^{-25}$ | $3.5 \pm 0.1 \times 10^{-25}$ | $3.2 \pm 0.1 \times 10^{-25}$ |
| | 22 | $6.7 \pm 0.2 \times 10^{-25}$ | $5.8 \pm 0.2 \times 10^{-25}$ | $6.5 \pm 0.1 \times 10^{-25}$ |
| | 55 | $2.7 \pm 0.2 \times 10^{-24}$ | $2.2 \pm 0.0 \times 10^{-24}$ | $1.9 \pm 0.1 \times 10^{-24}$ |
| dodecane | 10 | $3.6 \pm 0.2 \times 10^{-26}$ | $2.2 \pm 0.1 \times 10^{-26}$ | $1.8 \pm 0.1 \times 10^{-26}$ |
| | 22 | $7.3 \pm 0.2 \times 10^{-26}$ | $3.5 \pm 0.1 \times 10^{-26}$ | $3.1 \pm 0.3 \times 10^{-26}$ |
| | 55 | $2.0 \pm 0.0 \times 10^{-25}$ | $1.9 \pm 0.0 \times 10^{-25}$ | $1.4 \pm 0.0 \times 10^{-25}$ |

**Table 4 Calculated diffusion coefficients depending on *n*-alkane type, temperature, and layer thickness using fish gelatin as primary emulsifier.**

| *n*-alkane | temperature (°C) | primary o/w D ($m^2/s$) | secondary o/w D ($m^2/s$) | crosslinked o/w D ($m^2/s$) |
|---|---|---|---|---|
| octane | 10 | $2.0 \pm 0.1 \times 10^{-09}$ | $2.5 \pm 0.1 \times 10^{-09}$ | $1.6 \pm 0.1 \times 10^{-09}$ |
| | 22 | $3.1 \pm 0.1 \times 10^{-09}$ | $4.2 \pm 0.1 \times 10^{-09}$ | $2.3 \pm 0.1 \times 10^{-09}$ |
| | 55 | $6.2 \pm 0.3 \times 10^{-09}$ | $7.6 \pm 0.3 \times 10^{-09}$ | $6.9 \pm 0.1 \times 10^{-09}$ |
| decane | 10 | $1.6 \pm 0.0 \times 10^{-09}$ | $1.3 \pm 0.0 \times 10^{-09}$ | $1.2 \pm 0.0 \times 10^{-09}$ |
| | 22 | $2.7 \pm 0.1 \times 10^{-09}$ | $2.3 \pm 0.1 \times 10^{-09}$ | $2.5 \pm 0.0 \times 10^{-09}$ |
| | 55 | $1.2 \pm 0.1 \times 10^{-08}$ | $9.4 \pm 0.2 \times 10^{-09}$ | $8.1 \pm 0.2 \times 10^{-09}$ |
| dodecane | 10 | $1.5 \pm 0.1 \times 10^{-09}$ | $9.3 \pm 0.6 \times 10^{-10}$ | $7.7 \pm 0.5 \times 10^{-10}$ |
| | 22 | $3.2 \pm 0.1 \times 10^{-09}$ | $1.6 \pm 0.0 \times 10^{-09}$ | $1.4 \pm 0.1 \times 10^{-09}$ |
| | 55 | $9.8 \pm 0.1 \times 10^{-09}$ | $9.2 \pm 0.1 \times 10^{-09}$ | $6.9 \pm 0.1 \times 10^{-09}$ |

**Influence of n-alkane type on Ostwald ripening rate of primary emulsions**

Static light scattering was used to measure the changes in droplet size distribution over time for 0.5% (w/w) octane, decane, and dodecane oil-in-water emulsions stabilized by fish gelatin (**Figure 1**). The initial droplet diameter for the octane, decane, and dodecane emulsions were $d_{10} = 0.309 \pm 0.075$ µm, $0.170 \pm 0.009$ µm, and $0.120 \pm 0.001$ µm, respectively. The results noticeably show that the molecular structure of the dispersed phase in the continuous phase has a major impact on the molecular diffusion

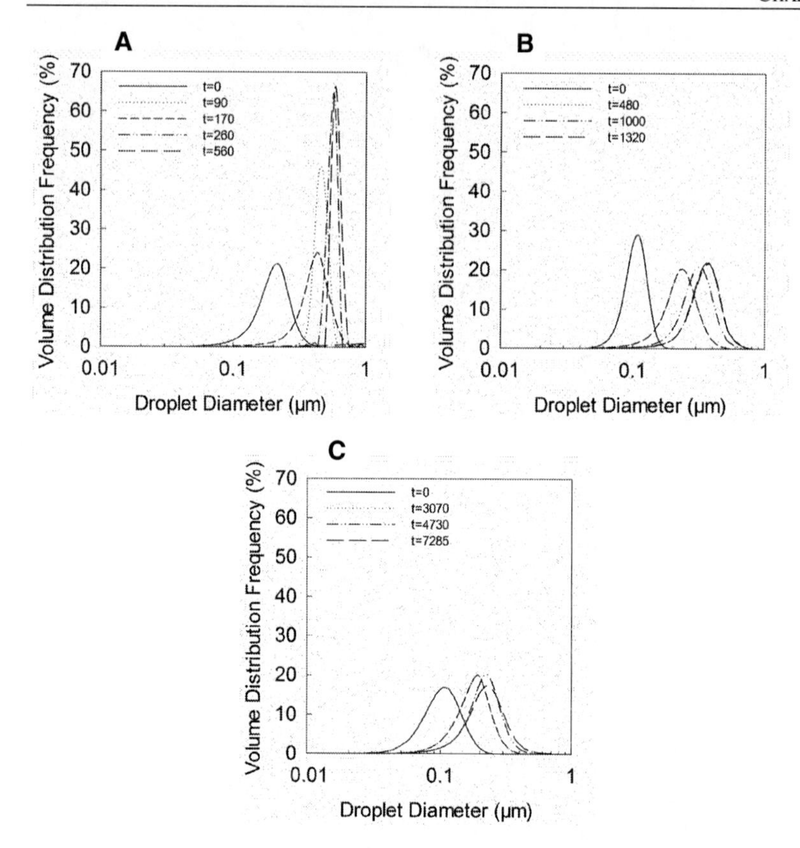


**Figure 1 Time dependence of particle size distributions of 0.5% (w/w) *n*-alkane emulsions stabilized by fish gelatin: octane (A), decane (B), dodecane (C). Droplet growth indicates Ostwald ripening (t in min).**

which supports the predictions of Equation (2) and is in line with previous studies (*Dickinson et al.*, 1999; *Weiss et al.*, 1999). The solubility of *n*-alkanes in the aqueous phase increases in the following order: octane > decane > dodecane, which enhances the mass transport between the droplets leading to a gradual shift of the monomodal size distribution to higher mean droplet sizes (**Figure 1**) (*Dickinson et al.*, 1999). In agreement with the Lifshitz-Slyozov-Wagner theory, there is a linear dependence of the normalized cube of the number-average radius $r^3$ on storage time. The Ostwald ripening rates were determined from the slope of the normalized $r_t^3$-$r_0^3$ versus $t$ plot using linear regression and given in **Table 3** (*Weiss et al.*, 1999). The growth rates from the linear regression analysis are: $\omega = 8.9 \pm 0.2 \times 10^{-24}$ m$^3$/s for octane, $6.7 \pm 0.2 \times 10^{-25}$ m$^3$/s for decane, and $7.3 \pm 0.2 \times 10^{-26}$ m$^3$/s for dodecane at a temperature of 22 °C (**Figure 2**). These values vary from previously reported ripening rates due to differences in the emulsifiers used (*Soma et al.*, 1996; *Weiss et al.*, 2000; *Sakai et al.*,

2002). Many studies used small molecular weight surfactant-stabilized model systems to investigate the ripening kinetics, whereas fish gelatin was used in our study to stabilize the emulsions. Fish gelatin is a flexible random coil protein that forms much thicker interfacial membranes than surfactants which may per se reduce the diffusion of oil molecules through the membrane compared to that through a thin small molecular weight surfactant membrane (*McClements*, 2004; *McClements*, 2004). *Wooster et al.* (2008) already described that the nature of the emulsifier used considerable affects the formation and stabilization of emulsions against Ostwald ripening.

**Changes of droplet size with time in multilayered emulsions – Influence of layer thickness on Ostwald ripening**

In our previous work, we had identified optimum preparation conditions for the electrostatic deposition method required to prepare stable oil-in-water emulsions containing droplets with fish gelatin-beet pectin membranes without incurring flocculation or coalescence of particles (*Zeeb et al.*, 2011; *Zeeb et al.*, 2012). The enzyme laccase was added to the secondary emulsions to promote interfacial crosslinking of adsorbed pectin molecules. In addition, the viscosity of the aqueous phase did not increase after separating the particles and analyzing the continuous phase suggesting almost no excess pectin is present which might interfere mass transport of alkane molecules through the aqueous phase (Data not shown).

The changes in mean average diameter with time and temperature are shown in **Figure 3** for dodecane oil-in-water emulsion stabilized by an enzyme-treated interfacial membrane complex, whereas the measured Ostwald ripening rates from the linear regression analysis are listed in **Table 3**.

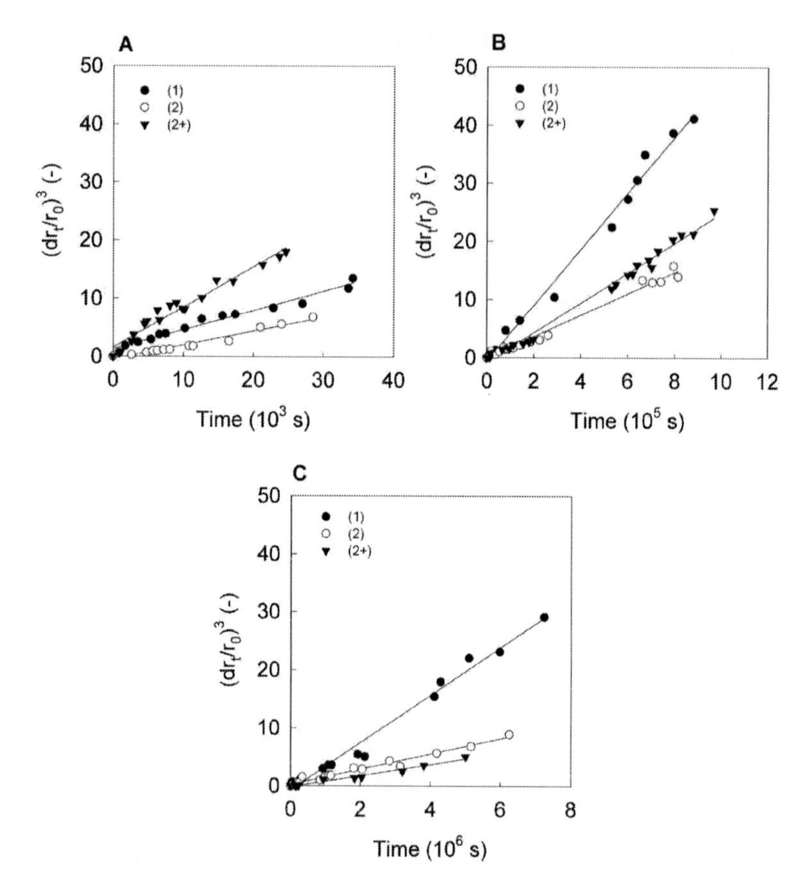


**Figure 2 Influence of layer thickness on molecular diffusion rates of octane (A), decane (B), and dodecane (C) oil-in-water emulsions at T = 22 °C.**

Our results indicate that Ostwald ripening in multilayered emulsions was still dominated by the molecular structure of the dispersed *n*-alkane phase. Growth rates of secondary and laccase-treated secondary emulsions increase as the water-solubility of the alkanes increases and normalized diffusion coefficients support these results (**Table 6**). Nevertheless, molecular diffusion could be retarded if thick and elastic interfacial membranes were built up (*Mun et al.*, 2006). Furthermore, *Mezzenga et al.* (2010) and *Adelmann et al.* (2012) provided a potential explanation for the observed reduction in Ostwald ripening, namely that crosslinked interfaces may slow down the tendency to change the shape of the particles due to an increased interfacial elasticity. This elasticity might be strong enough to preserve oil droplets, e.g. in spray-dried and freeze-dried samples. The rate of Ostwald ripening was appreciably lower for secondary and crosslinked secondary decane and dodecane emulsions than in emulsions where droplets had been stabilized only by fish gelatin (**Table 3**). In general, it is considered that crosslinking of adsorbed sugar beet

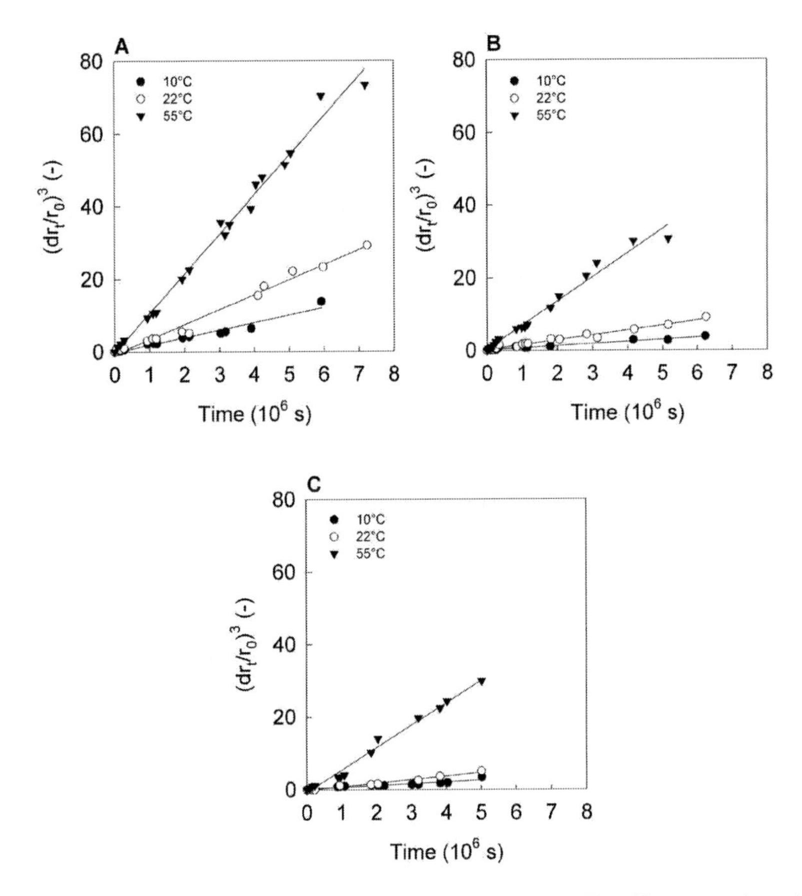


**Figure 3 Influence of temperature on Ostwald ripening of primary (A), secondary (B), and laccase-treated secondary (C) dodecane-in-water emulsions.**

pectin might alter the interactions between water and/or proteins and therefore change diffusion kinetics through the interfacial complex. However, the formation of multilayered membranes did not slow down molecular diffusion in octane-in-water emulsions (**Figure 2A**). This is not surprising since octane is firstly more soluble in water causing Ostwald ripening to generally proceed faster. By the time the emulsion can be coated, it may have already appreciably grown. Secondly, octane as a small molecule may diffuse more readily through the porous protein network. *Weers* (1998) already noted that the characteristic time $\tau_{OR} \approx r^3/\omega$ indicates the importance of Ostwald ripening in determining the initial droplet size of microemulsions. Droplets stabilized by SDS in decane-in-water emulsions ripen within a few minutes (73 min), whereas octane-in-water emulsions may grow even faster due to their higher water solubility ($c(\infty)_{octane} = 5.8 \times 10^{-3}$ mol/m$^3$; $c(\infty)_{decane} = 3.6 \times 10^{-4}$ mol/m$^3$) (*Soma et al.*, 1996; *Weers*, 1998). *Weers* (1998) published some characteristic time scales for Ostwald ripening in *n*-alkane-in-water emulsions stabilized by 0.1 M SDS. This explains why the initial number-average droplet size of the octane-in-water emulsions was almost twice in size in comparison of decane and dodecane oil-in-water emulsions.

A question that was raised and that at present we are unable to answer is whether "empty" protein shells remain as Ostwald ripening in crosslinked emulsions proceeds. Since there is a population of droplets that shrinks at the expense of a population of larger ones, one may hypothesize that the particles remain that consist of only protein shells. However, our light scattering studies did not reveal the presence of such a population, e.g. all measured particle size distributions were monomodal. Potentially, more insight could be gained from an experiment in which the system is separated into two fractions with one fraction containing particles of e.g. below 50 nm and one containing all other particles. The one containing smaller ones could then be subjected to additional dynamic light scattering measurements which might reveal the presence of such shells.

**Influence of temperature on Ostwald ripening of multilayered emulsions**

The purpose of these studies was to determine the influence of temperature on the Ostwald ripening kinetics of primary, secondary, and laccase-treated secondary emulsions. The effect of temperature on the changes in droplet size of *n*-alkane-in-water emulsions as a function of time is shown in **Figure 3**, whereas the growth rates are given in **Table 3**. The rates of molecular diffusion of multilayered emulsions showed a temperature-dependent behaviour. Droplet growth was accelerated due to faster diffusion of *n*-alkanes through the interfacial membranes and the aqueous phase, which was also described by *Delmas et al.* (2010). In this study, nanoemulsions stabilized by PEG-surfactants were prepared and exposed to different temperatures ranging from 25 to 70 °C. As previously mentioned, the diffusion of molecules is highly temperature dependent, a fact that may be deduced from Equation (3). Furthermore, it has been shown that laccase-induced crosslinking of adsorbed pectins in the interfacial membranes surrounding droplets in multilayered emulsions provides some limited improvement in terms of Ostwald ripening, in particular at elevated temperatures (55 °C).

**Influence of interfacial membrane density in secondary emulsions on Ostwald ripening**

The purpose of these studies was to determine the influence of membrane density on the Ostwald ripening kinetics of secondary dodecane-in-water emulsions. Dodecane was selected as a dispersed phase due to its low water solubility in comparison to octane or decane. Fish gelatin and WPI were chosen as emulsifiers to form sTable primary emulsions containing dodecane droplets which were subsequently coated with sugar beet pectin using electrostatic deposition. Fish gelatin is a random coil biopolymer having a highly dynamic and flexible structure, while WPI has a fairly rigid compact globular structure (*McClements*, 2004). Double-layered emulsions were heated to 80 °C and held for 10 min to promote a heat-induced crosslinking of the adsorbed protein layers (*Romoscanu et al.*, 2005). It is well established that most proteins such as WPI are prone to heat-induced conformational changes above approximately 60 °C which may involve unfolding of some segments followed by a sequential crosslinking due to free sulfhydryl or disulfide groups (*McClements et al.*, 1993; *Dickinson et al.*, 1994; *Rodríguez Patino et al.*, 1999). In comparison, fish gelatin undergoes a helix-coil transition forming a cold-set gel-like structure (*Walstra et al.*, 2003; *McClements*, 2004).

**Table 5 Measured Ostwald ripening rates (OR) and calculated diffusion coefficients (D) of heat-treated secondary dodecane-in-water emulsions (80 °C, 10 min) stabilized by a globular (WPI) or flexible (fish gelatin) protein-polysaccharide (beet pectin) membrane.**

| | protein layer | | | |
|---|---|---|---|---|
| **temperature** | **fish gelatin** | **fish gelatin heated** | **WPI** | **WPI heated** |
| **10 °C** | | | | |
| OR rate ($m^3/s$) | $2.2 \pm 0.1 \times 10^{-26}$ | $1.5 \pm 0.1 \times 10^{-26}$ | $7.9 \pm 0.0 \times 10^{-27}$ | $4.3 \pm 0.0 \times 10^{-27}$ |
| D ($m^2/s$) | $9.3 \pm 0.6 \times 10^{-10}$ | $6.4 \pm 0.2 \times 10^{-10}$ | $3.3 \pm 0.2 \times 10^{-10}$ | $1.8 \pm 0.2 \times 10^{-10}$ |
| **22 °C** | | | | |
| OR rate ($m^3/s$) | $3.5 \pm 0.1 \times 10^{-26}$ | $2.7 \pm 0.1 \times 10^{-26}$ | $1.1 \pm 0.0 \times 10^{-26}$ | $9.7 \pm 0.1 \times 10^{-27}$ |
| D ($m^2/s$) | $1.6 \pm 0.0 \times 10^{-09}$ | $1.2 \pm 0.0 \times 10^{-09}$ | $5.0 \pm 0.2 \times 10^{-10}$ | $4.3 \pm 0.3 \times 10^{-10}$ |
| **55 °C** | | | | |
| OR rate ($m^3/s$) | $1.9 \pm 0.0 \times 10^{-25}$ | $1.5 \pm 0.0 \times 10^{-25}$ | $6.3 \pm 0.0 \times 10^{-26}$ | $5.4 \pm 0.2 \times 10^{-26}$ |
| D ($m^2/s$) | $9.2 \pm 0.1 \times 10^{-09}$ | $7.5 \pm 0.1 \times 10^{-09}$ | $3.1 \pm 0.1 \times 10^{-09}$ | $2.6 \pm 0.1 \times 10^{-09}$ |

Our results indicated that the membrane density affects mass transport of dodecane molecules through the interfacial membrane. Ostwald ripening rates from the linear regression analysis are given in **Table 5**. Flexible random-coil biopolymers such as fish gelatin form a thick (~10 nm) but less dense interfacial membrane and are supposed to be more prone to Ostwald ripening. In fact, the ripening rates decreased from $3.5 \pm 0.1 \times 10^{-26}$ $m^3/s$ to $2.7 \pm 0.1 \times 10^{-26}$ $m^3/s$ after heat treatment at 22 °C. Possibly, a filamentous gel consisting of fish gelatin molecules is formed due to heating and subsequent cooling. Filamentous gels are also known to have a good water holding capacity because of the small pore sizes of the gel network where water is tightly held by capillary forces (*Belitz et al.*, 2001; *McClements*, 2004). Indeed, a replacement of fish gelatin by WPI led to a reduction in ripening rates, in particular after heat treatment of secondary emulsion. Ostwald ripening rates decreased from $1.1 \pm 0.0 \times 10^{-26}$ $m^3/s$ to $9.7 \pm 0.1 \times 10^{-27}$ $m^3/s$ after heat treatment at 22 °C (**Table 5**). This could be explained by the structure of the interfacial layer, although WPI provides a layer that only protrudes from the interface of an oil droplet to a depth of 1 to 2 nm (*Dalgleish et al.*, 1995). WPI as a globular protein is known to form dense and compact interfacial membranes, which retard Ostwald ripening better than fish gelatin (*Dalgleish et al.*, 1995; *McClements*, 2004). Even after heat-induced crosslinking of WPI molecules via disulfide bonds the interfacial membrane might be dense enough to slow down molecular diffusion of dodecane molecules (**Figure 4**). The rates of molecular diffusion of heat-treated and non heat-treated secondary emulsions also showed a temperature-dependent behaviour that could by described by Equation (3).

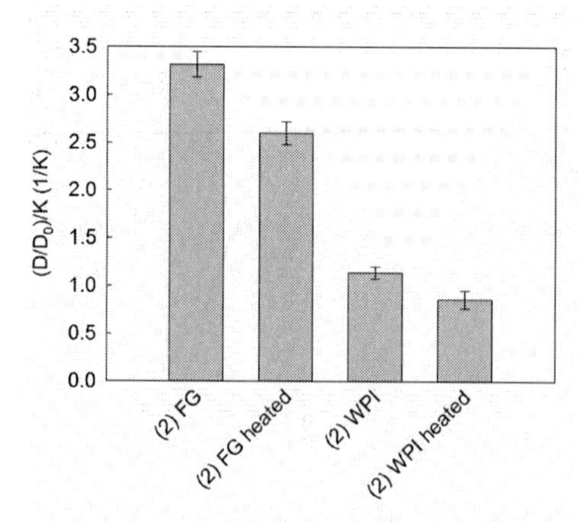


**Figure 4 Dependence of normalized diffusion coefficients ($D/D_0$)/T on temperature for secondary dodecane oil-in-water emulsions having different interfacial membrane densities.**

**Key insights**

In summary, we can highlight a number of key insights obtained by our study, as illustrated in **Figure 5**, where shorter arrows indicate a faster Ostwald ripening:

- The growth of droplets depends heavily on the molecular composition of the dispersed phase following in the order of their water solubilities: octane > decane > dodecane.
- The formation of multilayered emulsions using the layer-by-layer electrostatic deposition method increases the stability against Ostwald ripening. However, membranes of emulsions containing very short chain *n*-alkanes such as e.g. octane could not be coated and crosslinked fast enough to prevent molecular diffusion.
- A laccase-induced crosslinking of adsorbed sugar beet pectin in multilayered emulsions further enhanced the stability to Ostwald ripening, in particular in emulsions containing long chain *n*-alkanes due to the formation of a protective network.
- Heat-induced transformations retards Ostwald ripening in WPI-beet pectin coated and fish gelatin-beet pectin coated emulsions.

**Table 6 Normalized diffusion coefficients in dependence of temperature, layer thickness, and *n*-alkane.**

| *n*-alkane | temperature (°C) | primary o/w $D/D_0$ (-) | secondary o/w $D/D_0$ (-) | crosslinked o/w $D/D_0$ (-) |
|---|---|---|---|---|
| octane | 10 | 3.45 ± 0.21 | 4.36 ± 0.10 | 2.68 ± 0.14 |
| | 22 | 3.74 ± 0.10 | 5.04 ± 0.14 | 2.82 ± 0.08 |
| | 55 | 6.30 ± 0.40 | 7.06 ± 0.28 | 2.74 ± 0.14 |
| decane | 10 | 3.66 ± 0.06 | 3.02 ± 0.06 | 2.73 ± 0.07 |
| | 22 | 4.30 ± 0.10 | 3.61 ± 0.10 | 4.04 ± 0.07 |
| | 55 | 15.70 ± 1.09 | 12.61 ± 0.23 | 10.91 ± 0.31 |
| dodecane | 10 | 3.82 ± 0.17 | 2.33 ± 0.15 | 1.94 ± 0.12 |
| | 22 | 5.55 ± 0.15 | 2.74 ± 0.05 | 2.37 ± 0.19 |
| | 55 | 14.38 ± 0.21 | 13.59 ± 0.22 | 10.10 ± 0.31 |

- Ripening rates $\omega$ of secondary dodecane-in-water emulsions increase in the following order (at 22°C): heated WPI-pectin (2) < WPI-pectin (2) < heated fish gelatin-pectin (2) < fish gelatin-crosslinked pectin (2+) < fish gelatin-pectin (2) < fish gelatin (1).

- Generally, normalized diffusion coefficients showed a temperature-dependent behavior (**Table 6**). Ostwald ripening rates of multilayered emulsions increased with increasing temperatures.

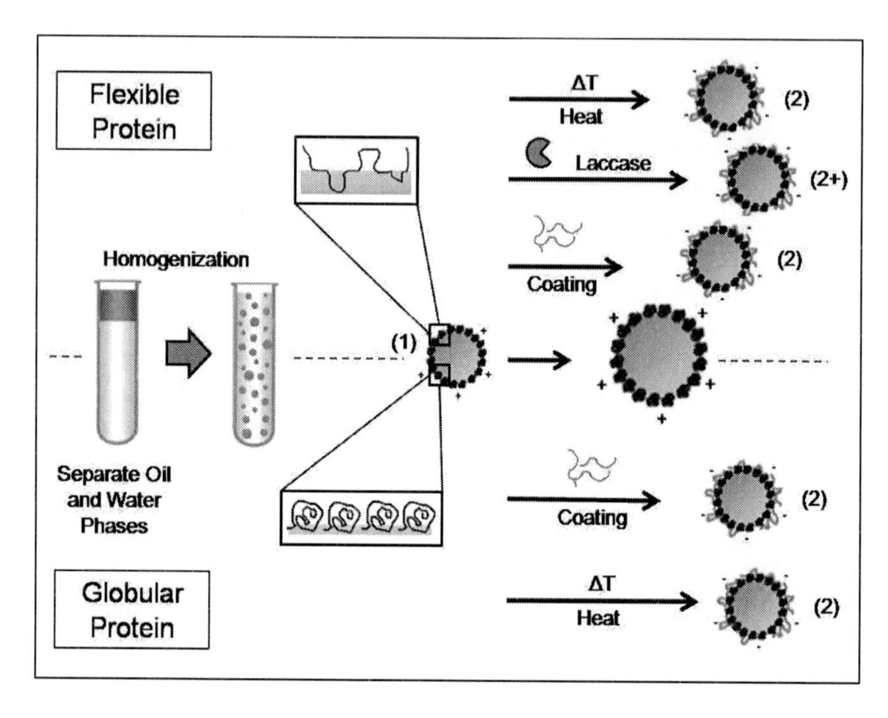


**Figure 5 Schematic mechanism of Ostwald ripening of primary (1), secondary (2), and laccase-treated secondary (2+) emulsions depending on interfacial thickness and density (shorter arrows indicate faster Ostwald ripening).**

## CONCLUSIONS

This study demonstrated that the rate of Ostwald ripening in *n*-alkane-in-water emulsions can be retarded by manipulating the properties of the interfacial membranes surrounding the oil droplets by the use of electrostatic depositioning of additional polymers. While the retardation effect is not as pronounced as the one that would be accomplished by exchanging the alkane with a higher molecular weight one, it nevertheless may prolong shelf life. We postulate that crosslinked adsorbed biopolymers form a network with small pore sizes that may slow down Ostwald ripening. In particular, in multilayered systems composed of both proteins and polysaccharides, a combination of physical and biomimetic-induced intramolecular crosslinking between one or more interfacial membranes and/or an entanglement of polymers may provide mechanical resistance to the shrinkage and growth of droplets or reduce pore size of the polymer network. This interfacial engineering approach may therefore be a useful tool for stabilizing emulsions against droplet growth and enabling manufactures to produce emulsions with improved physical stability during production, transport, and storage.

## ACKNOWLEDGEMENTS

We would like to thank Arla Foods Ingredients (Viby, Denmark) and Herbstreith & Fox KG (Neuenbürg, Germany) for generously providing us with samples.

## REFERENCES


Adelmann, H., Binks, B. P., et al. (2012): Oil powders and gels from particle-stabilized emulsions. Langmuir 28(3): 1694-1697.

Aoki, T., Decker, E. A., et al. (2005): Influence of environmental stresses on stability of O/W emulsions containing droplets stabilized by multilayered membranes produced by a layer-by-layer electrostatic deposition technique. Food Hydrocolloids 19(2): 209-220.

Belitz, H.-D., Grosch, W., et al. (2001): Lehrbuch der Lebensmittelchemie. Berlin, Springer-Verlag GmbH.

Dalgleish, D. G., Srinivasan, M., et al. (1995): Surface properties of oil-in-water emulsion droplets containing casein and tween 60. Journal of Agricultural and Food Chemistry 43(9): 2351-2355.

Davis, S. S., Round, H. P., et al. (1981): Ostwald ripening and the stability of emulsion systems: an explanation for the effect of an added third component. Journal of Colloid and Interface Science 80(2): 508-511.

De Smet, O. Y., Danino, D., et al. (2000): Ostwald ripening in the transient regime: A cryo-TEM study. Langmuir 16(3): 961-967.

Delmas, T., Piraux, H., et al. (2010): How to prepare and stabilize very small nanoemulsions. Langmuir 27(5): 1683-1692.

Dickinson, E. (1997): Enzymic crosslinking as a tool for food colloid rheology control and interfacial stabilization. Trends in Food Science and Technology 8(10): 334-339.

Dickinson, E., Golding, M. (1998): Influence of alcohol on stability of oil-in-water emulsions containing sodium caseinate. Journal of Colloid and Interface Science 197(1): 133-141.

Dickinson, E., Hong, S. T. (1994): Surface coverage of β-lactoglobulin at the oil-water interface: Influence of protein heat treatment and various emulsifiers. Journal of Agricultural and Food Chemistry 42(8): 1602-1606.

Dickinson, E., Ritzoulis, C., et al. (1999): Ostwald ripening of protein-stabilized emulsions: Effect of transglutaminase crosslinking. Colloids and Surfaces B: Biointerfaces 12(3-6): 139-146.

Gu, Y. S., Decker, A. E., et al. (2005): Production and characterization of oil-in-water emulsions containing droplets stabilized by multilayer membranes consisting of β-lactoglobulin, ι-carrageenan and gelatin. Langmuir 21(13): 5752-5760.

Gu, Y. S., Decker, E. A., et al. (2004): Influence of pH and ι-carrageenan concentration on physicochemical properties and stability of β-lactoglobulin-stabilized oil-in-water emulsions. Journal of Agricultural and Food Chemistry 52(11): 3626-3632.

Guzey, D., McClements, D. (2006): Influence of environmental stresses on o/w emulsions stabilized by β-lactoglobulin–pectin and β-lactoglobulin–pectin–chitosan membranes produced by the electrostatic layer-by-layer deposition technique. Food Biophysics 1(1): 30-40.

Guzey, D., McClements, D. J. (2006): Formation, stability and properties of multilayer emulsions for application in the food industry. Advances in Colloid and Interface Science 128-130: 227-248.

Hiemenz, P. C., Rajagopalan, R. (1997 ): Principles of colloid and surface chemistry, CRC Press.

Kabalnov, A. (2001): Ostwald ripening and related phenomena. Journal of Dispersion Science and Technology 22(1): 1-12.

Kabalnov, A. S., Pertzov, A. V., et al. (1987): Ostwald ripening in emulsions. I. Direct observations of Ostwald ripening in emulsions. Journal of Colloid and Interface Science 118(2): 590-597.

Kabalnov, A. S., Shchukin, E. D. (1992): Ostwald ripening theory: Applications to fluorocarbon emulsion stability. Advances in Colloid and Interface Science 38(C): 69-97.

Lifshitz, I. M., Slyozov, V. V. (1961): The kinetics of precipitation from supersaturated solid solutions. Journal of Physics and Chemistry of Solids 19(1-2): 35-50.

Littoz, F., McClements, D. J. (2008): Bio-mimetic approach to improving emulsion stability: Cross-linking adsorbed beet pectin layers using laccase. Food Hydrocolloids 22(7): 1203-1211.

Mamedov, A. M., Akhundov, T. S., et al. (1973): Dynamic viscosity of water, depending on temperature and pressure. Dinamicheskaya Vyazkost' Vody V Zavisimosti Ot Temperatury I Davleniya.(6): 72-74.

McClements, D. J. (2004): Food emulsions: Principles, practice, and techniques. Boca Raton, CRC Press.

McClements, D. J. (2004): Protein-stabilized emulsions. Current Opinion in Colloid & Interface Science 9(5): 305-313.

McClements, D. J., Decker, E. A., et al. (2009): Structural design principles for delivery of bioactive components in nutraceuticals and functional foods. Critical Reviews in Food Science and Nutrition 49(6): 577 - 606.

McClements, D. J., Li, Y. (2010): Structured emulsion-based delivery systems: Controlling the digestion and release of lipophilic food components. Advances in Colloid and Interface Science 159(2): 213-228.

McClements, D. J., Monahan, F. J., et al. (1993): Disulfide bond formation affects stability of whey protein isolate emulsions. Journal of Food Science 58(5): 1036-1039.

McClements, D. J., Stefan, K., et al. (2009). Biopolymers in food emulsions. Modern Biopolymer Science. San Diego, Academic Press: 129-166.

Meinders, M. B. J., Kloek, W., et al. (2001): Effect of surface elasticity on Ostwald ripening in emulsions. Langmuir 17(13): 3923-3929.

Meinders, M. B. J., van Vliet, T. (2004): The role of interfacial rheological properties on Ostwald ripening in emulsions. Advances in Colloid and Interface Science 108-109(0): 119-126.

Mezzenga, R., Ulrich, S. (2010): Spray-dried oil powder with ultrahigh oil content. Langmuir 26(22): 16658-16661.

Minussi, R. C., Pastore, G. M., et al. (2002): Potential applications of laccase in the food industry. Trends in Food Science & Technology 13(6-7): 205-216.

Mun, S., McClements, D. J. (2006): Influence of interfacial characteristics on Ostwald ripening in hydrocarbon oil-in-water emulsions. Langmuir 22(4): 1551-1554.

Ogawa, S., Decker, E. A., et al. (2003): Influence of environmental conditions on the stability of oil in water emulsions containing droplets stabilized by lecithin-chitosan membranes. Journal of Agricultural and Food Chemistry 51(18): 5522-5527.

Ogawa, S., Decker, E. A., et al. (2004): Production and characterization of O/W emulsions containing droplets stabilized by lecithin-chitosan-pectin multilayered membranes. Journal of Agricultural and Food Chemistry 52(11): 3595-3600.

Rodríguez Patino, J. M., Rodríguez Nino, M. R., et al. (1999): Dynamic interfacial rheology as a tool for the characterization of whey protein isolates gelation at the oil-water interface. Journal of Agricultural and Food Chemistry 47(9): 3640-3648.

Romoscanu, A. I., Mezzenga, R. (2005): Crosslinking and rheological characterization of adsorbed protein layers at the oil-water interface. Langmuir 21(21): 9689-9697.

Romoscanu, A. I., Mezzenga, R. (2006): Emulsion-templated fully reversible protein-in-oil gels. Langmuir 22(18): 7812-7818.

Sakai, T., Kamogawa, K., et al. (2002): Molecular diffusion of oil/water emulsions in surfactant-free conditions. Langmuir 18(6): 1985-1990.

Solans, C., Izquierdo, P., et al. (2005): Nano-emulsions. Current Opinion in Colloid & Interface Science 10(3-4): 102-110.

Soma, J., Papadopoulos, K. D. (1996): Ostwald ripening in sodium dodecyl sulfate-stabilized decane-in-water emulsions. Journal of Colloid and Interface Science 181(1): 225-231.

Taylor, P. (1998): Ostwald ripening in emulsions. Advances in Colloid and Interface Science 75(2): 107-163.

Voorhees, P. W. (1985): The theory of Ostwald ripening. Journal of Statistical Physics 38(1-2): 231-252.

Walstra, P., Walstra, W. (2003): Physical Chemistry of Foods, Marcel Dekker Inc.

Weers, J. G. (1998). Molecular diffusion in emulsions and emulsion mixtures. Modern Aspects of Emulsion Science, The Royal Society of Chemistry: 292-327.

Weiss, J., Canceliere, C., et al. (2000): Mass transport phenomena in oil-in-water emulsions containing surfactant micelles: Ostwald ripening. Langmuir 16(17): 6833-6838.

Weiss, J., Herrmann, N., et al. (1999): Ostwald ripening of hydrocarbon emulsion droplets in surfactant solutions. Langmuir 15(20): 6652-6657.

Wooster, T. J., Golding, M., et al. (2008): Impact of oil type on nanoemulsion formation and Ostwald ripening stability. Langmuir 24(22): 12758-12765.

Zeeb, B., Fischer, L., et al. (2011): Cross-linking of interfacial layers affects the salt and temperature stability of multilayered emulsions consisting of fish gelatin and sugar beet pectin. Journal of Agricultural and Food Chemistry 59(19): 10546-10555.

Zeeb, B., Gibis, M., et al. (2012): Crosslinking of interfacial layers in multilayered oil-in-water emulsions using laccase: Characterization and pH-stability. Food Hydrocolloids 27(1): 126-136.

# CHAPTER 5

## Influence of layer thickness and composition of crosslinked multilayered oil-in-water emulsions on the release behaviour of lutein

*Johanna Beicht*[1], *Benjamin Zeeb*[1], *Monika Gibis*[1], *Lutz Fischer*[2], *Jochen Weiss*[1]

[1] Department of Food Physics and Meat Science, University of Hohenheim, Garbenstrasse 21/25, 70599 Stuttgart, Germany

[2] Department of Food Biotechnology, University of Hohenheim, Garbenstrasse 25,70599 Stuttgart, Germany

Reproduced from "*Influence of layer thickness and composition of crosslinked multilayered oil-in-water emulsions on the release behavior of lutein*", Beicht, J., Zeeb, B., Gibis, M., Fischer, L., Weiss, J., Food Function, 2013, 4(10), p. 1457-1467 with permission from The Royal Society of Chemistry.

## ABSTRACT

Multilayering and enzymatic crosslinking of emulsions may cause alterations in the release behaviour of an encapsulated core material due to changes in thickness, porosity and permeability of the membrane. To test this hypothesis, an interfacial engineering technology based on the layer-by-layer electrostatic deposition of oppositely charged biopolymers onto the surfaces of emulsion droplets in combination with an enzymatic treatment was used to generate emulsions with different droplet interfaces. Release behaviour of primary, secondary (coated), and laccase-treated secondary emulsions carrying lutein, an oxygenated carotenoid, were characterized and studied. For our experiments fish gelatin (FG), whey protein isolate (WPI) and dodecyltrimethylammonium bromide (DTAB) as primary emulsifiers at acidic conditions (pH 3.5) was used to facilitate the adsorption of a negatively charged biopolymer (sugar beet pectin). Laccase was added to promote crosslinking of adsorbed beet pectin. The release of lutein-loaded emulsions was investigated and quantified by UV-vis-spectrophotometry. Primary WPI–stabilized emulsions showed a five times higher release of lutein after 48 h than secondary emulsions (pH 3.5). Primary DTAB-stabilized emulsions released 7.2% of encapsulated lutein within the observation period, whereas beet pectin-DTAB coated emulsions released only 0.13% of lutein. Crosslinking of adsorbed pectin did not significantly decrease release of lutein in comparison to non-crosslinked secondary emulsions. Additionally, release of lutein was also affected by changes in the pH of the surrounding medium. Results suggest that modulating the interfacial properties of oil-in-water emulsion by biopolymer deposition and/or crosslinking may be a useful approach to generate food grade delivery systems that have specific release-over-time profiles of incorporated active ingredients.


**Keywords:** Release; Lutein; Multilayered oil-in-water emulsions; Laccase; Sugar beet pectin; Fish gelatin; DTAB; Layer-by-layer depositioning

## INTRODUCTION

There has been a growing interest of consumers, food manufacturers, and scientists to functionalize food products. Instead of simply providing sufficient quantities of safe food required to satisfy the caloric intake needs of the population, consumer demands increasingly ask that their food not only provide calories but promote as well their health and wellbeing (*Shefer et al.*, 2003). The appearance of food ingredients with health and wellness benefits offers an excellent chance to improve consumers' health (*Chen et al.*, 2006).These food products are known as nutraceuticals and functional foods. The list of bioactive or nutraceutical components is endless. Carotenoids, omega-3 fatty acids, vitamins, bioactive peptides, antioxidants, minerals, modified fats, and flavors are only a few with health-promoting effects. Lutein is one of the best-known carotenoids which could naturally be found as lipid soluble pigment in various vegetables (e.g. spinach, kale) and is commercially extracted of marigold (*Tagetes*) (*Sowbhagya et al.*, 2004) (**Figure 1**). With the growing legislative restrictions on the use of synthetic colors, naturally colorants have received an upcoming attention within the food industry. According to new EU regulations lutein E 161b can be used as a coloring agent in foods (EU-no. 1129/2011). Moreover, *Kline et al.* (2011) demonstrated that lutein provides not only coloring properties, but could also reduce the formation of hexanal off-flavors in a colloidal beverage system suggesting implications as a photoprotective agent fortification. Recent studies showed that consumption of lutein may lower the incidence of age-related macular degeneration (AMD) and cataracts, the leading causes of acquired blindness in the elderly population (*Barker Ii*, 2010).

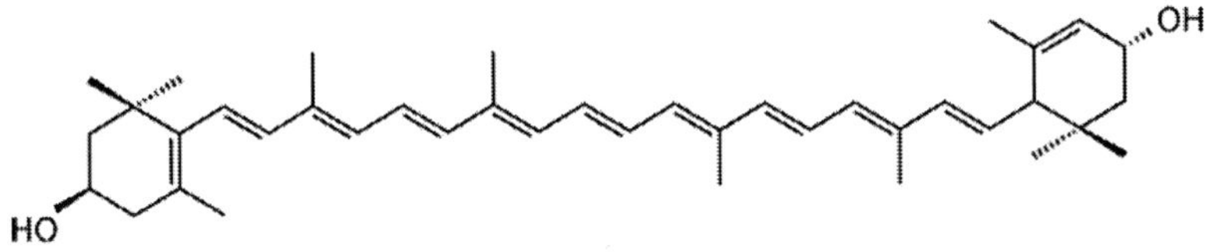

HO OH

**Figure 1 Chemical structure of lutein.**

A major challenge of food technologists is to incorporate these bioactive ingredients such as lutein into the food matrix preserving their bioavailability with specific release behaviours (*Li et al.*, 2010; *Ahmed et al.*, 2012). In particular, *Vishwanathan et al.* (2009) for example demonstrated that nanoemulsion-based delivery systems of lutein had significantly higher bioavailability than supplement-pills. Based on emulsion technologies, a wide variety of different delivery systems have been developed to encapsulate lipophilic bioactive components with specific functionalities such as efficient encapsulation and protection against degradation, compatibility within the food matrix, resistance to environmental stresses and specific release behaviors (*McClements et al.*, 2007; *McClements et al.*, 2009; *McClements et al.*, 2010; *Matalanis et al.*, 2011). The design of delivery systems having controlled release mechanisms such as delayed release in the colon for drugs or sustained release for flavours e.g. in chewing gums need to be created for each application (*Kosaraju*, 2005; *Madene et al.*, 2006).

Proteins or low-molecular surfactants are not only used to stabilize oil-in-water emulsions, but also to manipulate their interfacial properties since they differ in their chemical structure, their ability to reduce the surface tension, their network formation, and their conformational changes at interfaces (*Bos et al.*, 2001; *Wilde et al.*, 2004). A number of previous studies examined the impact of initial emulsifier type on the in vitro digestion models (*McClements et al.*, 2010; *Nik et al.*, 2012). However, many oil-in-water emulsions are prone to physical instabilities due to environmental stresses such as high temperatures, pH adjustments, and exposure to pressure or shear. A well-established interfacial engineering technology based on the layer-by-layer depositioning in combination with an enzymatic treatment was developed to overcome this deficit to generate stable emulsions with specific interfacial properties. The layer-by-layer technique is based on the electrostatic deposition of oppositely charged polyelectrolyte layers onto charged surfaces of emulsion droplets (*Guzey et al.*, 2006). Biopolymer coatings of droplets are already known to increase the emulsions stability against environmental stresses (*Moreau et al.*, 2003; *Aoki et al.*, 2005; *Klinkesorn et al.*, 2005). Moreover, multilamellar layers allowed to specifically release an encapsulated component due to a trigger mechanism (*Benjamin et al.*, 2012). The functional properties of these coatings can be further modified by enzymatic crosslinking of the adsorbed biopolymers. Recent studies have previously shown that laccase-oxidized ferulic acid groups in sugar beet pectin to from crosslinked pectin molecules at an oil droplet surfaces (*Littoz et al.*, 2008; *Zeeb et al.*, 2011; *Jung et al.*, 2012; *Zeeb et al.*, 2012; *Zeeb et al.*, 2012; *Zaidel et al.*, 2013). Crosslinked pectin coatings showed increased stability and higher resistance against environmental stresses such as pH, temperature or salt addition (*Jung et al.*, 2012; *Zeeb et al.*, 2012). However, *Sandra et al.* (2008) demonstrated that interfacial crosslinking of globular protein did not influence the digestion of emulsified lipids under simulated human small intestine conditions. The diffusion of the core material in the surrounded medium is dependent on the interfacial properties of the emulsion and may be manipulated by layer thickness, porosity, permeability, and enzymatic modifications. Furthermore, the susceptibility to environmental stresses may act as a specific trigger to release the bioactive ingredients at the right side of action.

The objective of the present study was to investigate the effect of membrane composition and treatment on the release behavior of multilayered oil-in-water emulsions. To this purpose, we studied the release of lutein from primary, secondary (coated), and laccase-treated secondary emulsions. We hypothesized that modulating the interfacial properties of emulsions by biopolymer deposition and enzymatic treatment may generate delivery systems with specific release-over-time profiles. Oil-in-water emulsions were therefore fabricated with a single- or double-layered membrane, treated with laccase, and subjected to various pH-values. To study the influence of interface composition on the release kinetics of lutein, we used three different kinds of emulsifiers, namely fish gelatin (FG), whey protein isolate (WPI), and dodecyltrimethylammonium bromide (DTAB). All of these primary emulsifiers could be coated with negatively charged sugar beet pectin at low pH values.

## MATERIALS AND METHODS

**Materials.** Whey protein isolate (WPI, #B180214) was purchased from Arla Foods Ingredients (Viby, Denmark). As per manufacturers specifications, WPI contained $\geq$ 88% protein, $\leq$ 6.0% moisture, $\leq$ 4.5% ash, $\leq$ 0.2% lactose, and $\leq$ 0.2% fat. WPI was used without further purification. Cold water fish skin gelatin (FG, #049K0050) and dodecyltrimethylammonium bromide (DTAB, #D8638) were obtained from Sigma-Aldrich GmbH (Steinheim, Germany). Sugar beet pectin (Pectin Betapec RU 301, #10903135) was purchased from Herbstreith & Fox KG (Neuenbürg, Germany). As stated by the manufacturer the degree of esterification was 55%. Laccase (#0001437590, from *Trametes versicolor*) was obtained from Sigma–Aldrich Co. (Steinheim, Germany). The laccase was reported to have 13.6 Units per mg (AU) of enzyme. Miglyol 812N, a medium chain triacylglyceride mixture, was purchased from Sassol Germany GmbH (Brunsbüttel, Germany). Marigold Oleoresin (Lutein E 161 b) was kindly supplied by Rudolf Wild GmbH & Co. KG (Heidelberg/Eppelheim, Germany). As stated by the manufacturers specification, oleoresin was obtained by solvent extraction of dried Marigold flowers of *Tergetes erecta* and had a xanthopyll content of ≥ 112 g/kg. Butylatedhydroxytoluene (BHT, #MKBC6147, purity ≥ 99% FCC, kosher) was obtained by SAFC (St. Louis, MO). Tri-sodium citrate dihydrate (#3580.1, purity ≥ 99%), citric acid monohydrate (#39581, purity ≥ 99.5%), analytical grade hydrochloric acid (HCl), and sodium hydroxide (NaOH) were obtained from Carl Roth GmbH & Co. KG (Karlsruhe, Germany). Distilled water was used for the preparation of all samples.

**Solution Preparation.** Aqueous FG-, WPI- and DTAB-solutions were prepared by dispersing 1% (w/w) of emulsifier powder into 10 mM citrate buffer (pH 3.5) and stirred for at least 2 h. Sodium azide solution (0.02% (w/w)) was added to the protein solutions to prevent microbial growth. Sugar beet pectin solution was prepared by dispersing 2% (w/w) powdered pectin into 10 mM citrate buffer at pH 3.5 followed by stirring overnight to ensure complete hydration. pH was then adjusted to 3.5 using 1 M HCl and/or 1M NaOH. Enzyme solution was prepared by dispersing enzyme powder into 10 mM citrate buffer (pH 3.5) followed by stirring for 30 min. Lutein was dissolved in Miglyol and sonicated in an ultrasound bath for at least 2 h. Prior lutein solubilization, BHT was dissolved in Miglyol (0.1% (w/w)) to prevent oxidation of lutein.

**Preparation of lutein-loaded primary, secondary, and enzyme-treated emulsions.** A stock emulsion was prepared by mixing 10 g of oil with 90 g of emulsifier solution (1% (w/w) 10 mM, in citrate buffer, pH 3.5)) in a glass beaker to obtain a 10% (w/w) primary oil-in-water emulsion. The oil phase contained different concentrations of lutein to determine the maximum loading capacity (1%, 3%, 5%, 20% and 30% w/w). Oil and emulsifier solutions were blended using a high shear blender (Standard Unit, IKA Werk GmbH, Germany) for 2 min and then passed through a high-pressure homogenizer (M110-EH-30, Microfluidics International Cooperation, Newton, MA) three times at 10000 psi (= 68.95 MPa). Secondary emulsions were prepared by mixing primary emulsions with beet pectin solution and 10 mM citrate buffer (pH 3.5) to yield emulsions with the following compositions: 1% (v/v) oil, 0.09% (v/v) WPI, FG or DTAB, and 0.2% (v/v) pectin to coat WPI- and FG-stabilized emulsions and 0.1% (v/v) pectin to coat DTAB-stabilized emulsions. Laccase-treated secondary emulsions were prepared as previously described by *Zeeb et al.* (2012). An enzyme/beet pectin ratio of 0.24 mg/4 mg, was sufficient to promote interfacial crosslinking of sugar beet pectin.

**Particle size determination.** Particle size distributions and polydispersity indices were measured using a dynamic light scattering instrument (Nano ZS, Malvern Instruments,

Malvern, U.K.). Prior analysis, emulsions were diluted to a droplet concentration of approximately 0.005% (v/v) with an appropriate buffer to avoid multiple scattering effects. The instrument calculates the particle diameter by determining the time-dependence of the intensity of scattered light from oil droplets that move in the aqueous phase due to Brownian motion. The size is then calculated from the diffusion constant using the Einstein equation (*Dalgleish et al.*, 1995). The instrument reports the mean particle diameter (z-average) and the polydispersity index (PDI) ranging from 0 (monodisperse) to 0.50 (very broad distribution). Every measurement was made with three readings per sample.

**ζ-Potential measurements.** Emulsions were diluted to a droplet concentration of approximately 0.005% (v/v) using an appropriate buffer solution. Diluted emulsions were then injected into a cuvette of a particle electrophoresis instrument (Nano, ZS, Malvern Instruments, Malvern, U.K.). The emulsions droplet electric potential (ζ-potential) was determined by measuring the direction and velocity that the droplets moved in the applied electric field and calculated using the Smoluchowski equation.

**Color measurements.** *L*, a*, b** values in the CIE tristimulus color space were measured using a chromameter CR-200 (Konica Minolta, Marunouichu Japan) to determine the optical properties of the samples. The chromameter was calibrated against a white calibration plate with tristimulus values: Y=93.7, x=0.3159, y=0.3321 at $D_{65}$ illumination. A fixed amount of emulsion sample was poured into a measurement cup, which was covered with a white tile before the measurement was carried out. Each emulsion sample was gently mixed before measurement to ensure a uniform particle distribution. The instrument provides the color of the samples in terms of *L*, *a*, and *b* color space system (*McClements*, 2004). In this color space, *L* represents the "lightness" and *a* and *b* are color coordinates: +*a* is the red direction, -*a* is the green direction, +*b* is the yellow direction, and –*b* is the blue direction (*McClements*, 2002).

**Optical microscopy.** The microstructure of all emulsions was investigated by optical microscopy. Each emulsion sample was gently mixed before analysis. One drop of emulsion was placed on an objective slide and then covered with a coverslip. Light microscopy images were taken with an axial mounted Canon Powershot G10 digital camera (Canon, Tokyo, Japan) mounted on an Axio Scope optical microscope (A1, Carl Zeiss Microimaging GmbH, Göttingen, Germany) at resolution of 40 x.

**Release of lutein into Miglyol as acceptor medium.** The influence of interfacial composition and treatment on the release behaviour of lutein-loaded primary, secondary, and enzyme-treated secondary emulsions was investigated. A modified membrane free model system was used to induce the release of lutein (*Mitri et al.*, 2011). Lutein-loaded primary, secondary and, enzyme-treated emulsions (1% (v/v) oil droplet concentration) were therefore transferred into a 100 mL glass beaker (20 ml) and Miglyol as release medium was carefully placed on top of each emulsion (20 ml). The release experiment was performed at 350 rpm (using a cylindrically shaped magnetic stirrer bar; PTFE; 25 x 8 mm) on a stirrer plate to avoid mixing of both liquid phases at ambient temperature. By stirring the lower but not the upper phase the lutein concentration in the water was equilibrated and transfer to the lipid acceptor phase occurred very rapidly due to low diffusion pathways. Aliquots were carefully taken from the oil phase at regular time intervals (700 µL) and directly replaced with fresh acceptor medium to maintain sink conditions. UV-visible measurements were used to determine the release of lutein into the acceptor medium. The release of lutein was followed by measuring the increase in absorbance at 453 nm using a UV-spectrophotometer (LAMBDA 750s UV/Vis/NIR Spectrophotometer, Perkin Elmer, Shelton, USA). Quantification of lutein in Miglyol was performed using a calibration curve (0.78, 1.96, 3.91, 5.87, 7.83, 9.78, 11.75 mg/l lutein) obtained with lutein standards.

**pH-dependent release behaviour of lutein.** The release behaviour of lutein-loaded primary, secondary, and enzyme-treated secondary emulsions under different pH values was also determined. Therefore, all emulsion samples were adjusted to different pH values prior the release study. Subsequent release experiments were performed as previously described.

**Statistical analysis.** All measurements were repeated at least twice using duplicate samples. Average and standard deviations were calculated from these duplicate measurements using Excel (Microsoft, Redmond, WA). Results were analyzed with statistical software (SAS 9.2, SAS Institute, Cary, NC). Variance analysis was performed using the Sidak- and Scheffe test for unbalanced and normally distributed data ($\alpha = 0.05$). If normally distributed results were in doubt, a nonparametric test according to Kruskal-Wallis was performed.

## RESULTS AND DISCUSSION

### Characterization of primary base oil-in-water emulsions

Release behaviour of bioactive core materials is affected by the interfacial properties of the membrane complex such as thickness, porosity, and permeability. Therefore, three different emulsifier types were used to stabilize base oil-in-water emulsions (1% (w/w) emulsifier solution): WPI, FG, and DTAB, respectively. The emulsifiers used vary in their chemical structure, charge, and molecular size, and therefore may show differences in releasing lutein into the surrounding phase. DTAB is a small, cationic surfactant, which forms a thin layer surrounding the oil droplets. In contrary, fish gelatin is a random coil biopolymer having a highly dynamic and flexible structure, while WPI has a fairly rigid compact globular structure. Both proteins provide much thicker membranes (~ 2 to 10 nm), but vary significantly in their interfacial density due to their chemical composition. First, drug free oil-in-water emulsions stabilized by DTAB, FG, and WPI (10% (w/w) Miglyol) were produced under constant homogenization conditions (10000 psi, 3 passes). Particle size distribution, ζ-potential, and microstructure of freshly prepared samples were assessed immediately after homogenization (**Table 1**). Our results show that all emulsifiers may be used to form stable small-sized oil-in-water emulsions containing positive charged droplets. In addition, the polydispersity index of all emulsions was 0.15 ± 0.02 indicating a very narrow particle size distribution – an important fact in producing multilamellar membrane coatings (*Zeeb et al.*, 2012). In particular, WPI is well known as a suitable emulsifier at acidic conditions since the protein is highly positively charged below its p*I* and therefore forms a protective interfacial membrane (*Demetriades et al.*, 1997). Additionally, steric repulsion between the WPI-stabilized oil droplets maintain the overall stability of the emulsion under prevalent conditions (*Malaki Nik et al.*, 2010). Moreover, FG has previously been shown to act as a suitable emulsifier at acidic conditions (*Zeeb et al.*, 2011; *Zeeb et al.*, 2012). In comparison to both proteins, low molecular weight surfactants such as DTAB are more effective in reducing the interfacial tension between the oil and water phase, thus resulting in smaller particle sizes during homogenization as shown in **Table 1**. However, surfactant-stabilized oil-in-water emulsions are often more prone to coalescence due to the fact that interfacial membranes formed by surfactants are thinner without providing a steric barrier in comparison to globular or flexible proteins (*Bos et al.*, 2001; *McClements*, 2004; *Wilde et al.*, 2004).

**Table 1 Characteristics of base emulsions containing droplets stabilized by fish gelatin (FG), whey protein isolate (WPI), or DTAB membranes.**

| emulsifier type | mean particle diameter (nm) | polydispersity index (-) | ζ-potential (mV) |
|---|---|---|---|
| WPI | 202 ± 4 | 0.15 ± 0.02 | 28.0 ±1.5 |
| FG | 200 ± 6 | 0.16 ± 0.02 | 21.0 ±1.0 |
| DTAB | 138 ± 10 | 0.15 ± 0.02 | 45.0 ± 5.2 |

**Determination of loading capacity of single-layered oil-in-water emulsions**

FG-, WPI-, and DTAB-stabilized oil-in-water emulsions loaded with various concentrations of lutein (1%, 3%, 5%, 20%, 30% (w/w) loading per 10% carrier oil) were produced by microfluidization to determine the maximum loading capacity of each carrier system. We examined the mean particle diameter, ζ-potential, and creaming stability of the freshly prepared primary emulsions. Our results indicate that the loading concentration of lutein affects the droplet diameter distributions of all base emulsions (**Figure 2**). We could prepare stable WPI-stabilized emulsions having a monomodal particle size distribution (PDI 0.18 ± 0.05) and a mean droplet diameter of 203.0 ± 3.5 nm with 5% (w/w) lutein-loading (**Figure 2B**). WPI–stabilized oil-in-water emulsions have previously shown to be suitable as a delivery system of polyunsaturated lipids, such as *w*-3 fatty acids and β-carotene (*Djordjevic et al.*, 2004; *Like et al.*, 2010). Higher lutein-loadings up to 20% (w/w) led to an increase in particle size and a formation of a creamed emulsion layer could be observed after > 7 days (**Figure 2B**). In general, increasing lutein concentrations may change the physical properties of the oil phase, in particular the viscosity, having a tremendous impact on the formation of stable emulsions as previously described by *Wooster et al.* (2008). In their study, it was shown that nanoemulsions made with high viscosity oils were considerably larger than nanoemulsions prepared with low viscosity oils. Moreover, it was shown that the mean particle diameter of FG-stabilized emulsions increased with increasing concentrations of encapsulated lutein, whereas the PDI increased from 0.06 ± 0.03 to 0.22 ± 0.06 and a creamed layer became visible for all loading concentrations used in our study immediately after homogenization (**Figure 2A** and **C**). FG is known to have a relatively low surface activity and is therefore less suitable in stabilization of emulsions. *Surh et al.* (2006) previously observed the formation of large droplets (>10 μm) in fish gelatin-stabilized emulsions.

Apart from FG and WPI, DTAB was used to produce surfactant-stabilized emulsions forming a much thinner interfacial membranes. We could prepare stable DTAB-stabilized emulsions loaded with 5% lutein (10% (w/w) carrier oil). However, buffer-diluted DTAB-stabilized emulsions were considerably more prone to aggregation than emulsions diluted with a DTAB-solution (0.5% (w/w)). In particular, under shearing conditions similar to those occurring during the release experiments, the buffer-diluted emulsion was even more prone to break. These results are in line with previous studies described by *Georgieva et al.* (2009), where DTAB-stabilized emulsions showed a lower physical stability compared to nonionic surfactants–stabilized emulsions. A dilution with buffer may lead to higher desorption of surfactant molecules from the surface into the bulk phase, which might induce coalescence after collision of the droplets and most recently the breakdown of the emulsion itself.

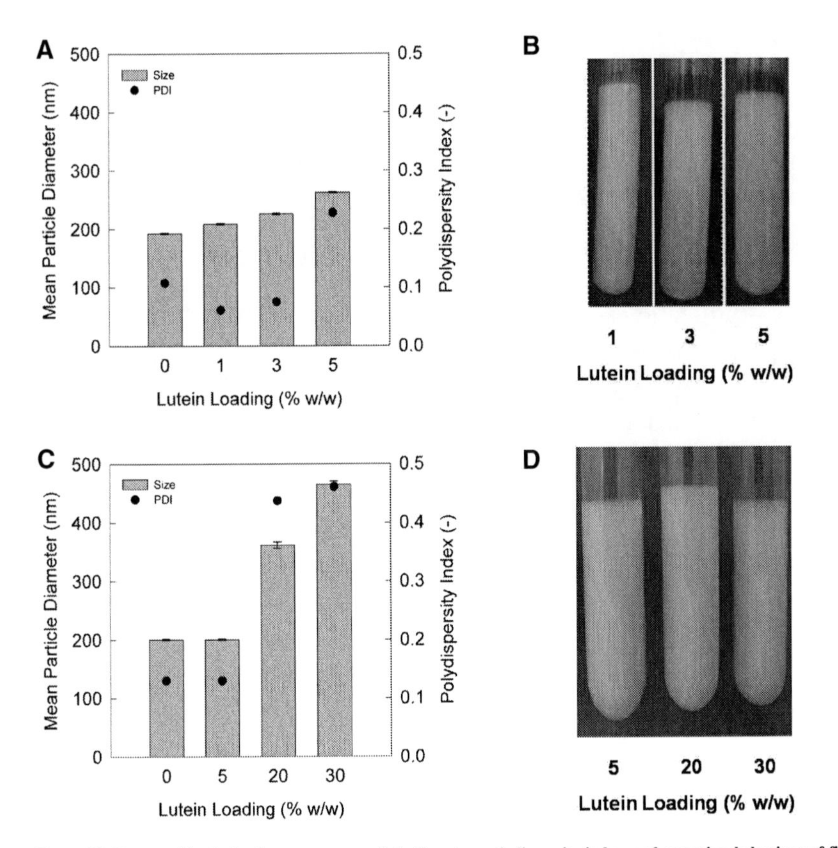


**Figure 2 Influence of lutein loading on mean particle diameter, polydispersity index, and creaming behaviour of fish gelatin (A, C) and WPI (B, D) stabilized oil-in-water emulsions (10% (w/w) oil, 10mM citrate buffer, pH 3.5).**

The optimum lutein loading reqiured to form stable, non-aggregated primary base emulsions was 5% (per 10% (w/w) oil) and therefore this loading level was used for all subsequent experminets. Moreover, primary DTAB-stabilized emulsions were diluted with a 0.5% (w/v) DTAB-solution for all release studies to prevent coalescences. We did not use FG for any release studies, since we could not prepare stable lutein-loaded emulsions stabilized by FG membranes for any loading concentration.

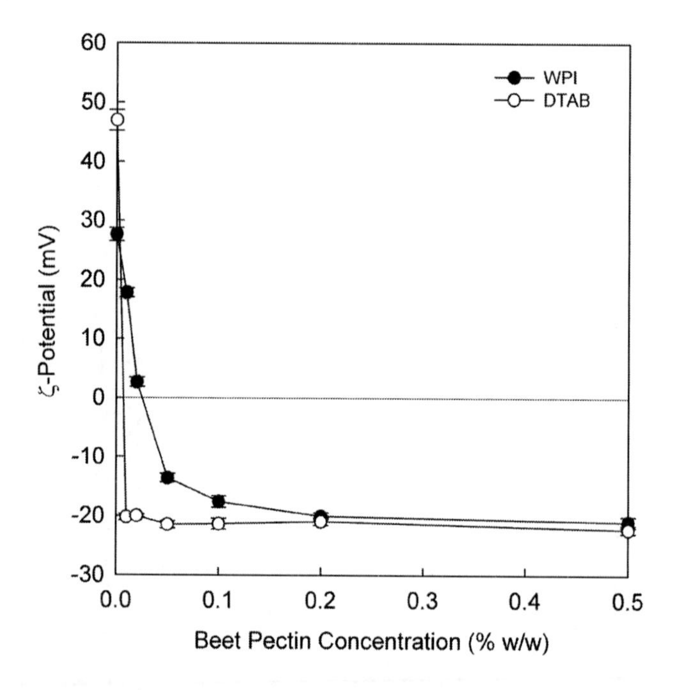


**Figure 3 Particle charge (ζ-potential) of WPI- and DTAB-stabilized oil-in-water emulsions (1% w/w Miglyol, 10 mM citrate buffer, pH 3.5) as a function of added sugar beet pectin (0 - 0.5% (w/v) oil).**

## Preparation of secondary and enzyme-treated emulsions carrying lutein as bioactive material

An interfacial engineering technology based on the layer-by-layer electrostatic deposition of oppositely charged biopolymers was used to form double-layered emulsions (*Guzey et al.*, 2006). The change in electrical charge of emulsions (1% (w/w) oil) to which different concentrations of sugar beet pectin (0 – 0.5%) had been added was measured (**Figure 3**). In the absence of pectin, the electrical charge of WPI-stabilized emulsion was +28.0 ± 1.1 mV, whereas DTAB-stabilized emulsion showed a significantly higher positive charge of +45.0 ±

5.2 mV (citrate buffer, pH 3.5). The electrical charge on the WPI-stabilized droplets became increasingly negative as pectin was added to the emulsion suggesting that the negatively charged pectin adsorbed to the surface of the positively charged droplets forming a WPI-pectin membrane (**Figure 3**). The ζ-potential became constant at a value of -20.0 ± 1.2 mV when the pectin concentration exceed about 0.2% (w/v) which indicates that the droplets became fully saturated with beet pectin (**Figure 3**). When pectin was added to DTAB-stabilized emulsion, the ζ-potential immediately went from positive to negative (**Figure 3**). However, stable double-layered emulsions could be prepared when the pectin concentration exceeded 0.1% (w/v). This may be attributed to differences in surface charge densities between the interfacial membranes of DTAB and WPI leading to a stronger adsorption of pectin molecules. The mean particle diameter of primary WPI-stabilized emulsions increased from 202 ± 4 nm to 346 ± 8 nm, whereas DTAB-stabilized emulsions increased from 138 ± 10 nm to 321 ± 5 nm after pectin was adsorbed.

**Storage stability of primary, secondary and enzyme treated lutein-loaded emulsion**

Release behaviour of core material is affected by the physical stability of the emulsion, therefore particle size distribution, ζ-potential, and creaming behaviour of primary, secondary and enzyme-treated emulsions (5% (w/w) lutein per 1% (w/w) oil) were investigated over a storage time of 9 days. The mean particle diameter and ζ-potential remained unchanged as well as the polydispersity indices of all emulsion samples (0.22 ± 0.07) over the entire storage time (**Figure 4**). Moreover, photographic images did not show any creamed layer (Data not shown). Citral-loaded primary emulsion stabilized by lecithin showed no obvious creaming or phase separation (*Yang et al.*, 2012). Moreover, secondary emulsions have previously been shown to have an improved stability to droplet aggregation because of relatively strong steric repulsive forces that are associated with the thick interfacial membrane complex (*Guzey et al.*, 2004; *Aoki et al.*, 2005; *Gu et al.*, 2005; *Zeeb et al.*, 2012). In particular, multilayered emulsions as delivery systems for citral have been demonstrated to increase the physical stability against creaming (*Yang et al.*, 2012).

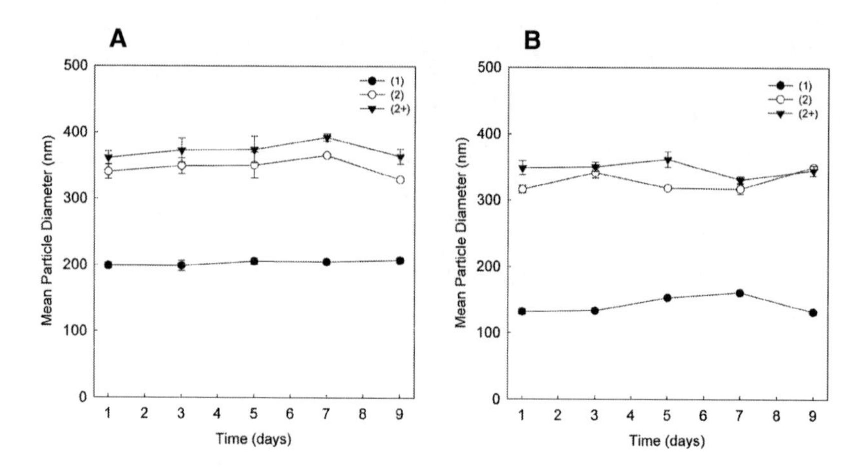


**Figure 4 Mean particle diameter of lutein-loaded WPI-stabilized (A) and DTAB-stabilized (B) primary (1), secondary (2), and laccase-treated, secondary emulsions (2+) as a function of time (5% (w/w) lutein-loading, 10 mM citrate buffer, pH 3.5).**

Citral-loaded primary emulsion stabilized by lecithin and two-layered emulsions consisting of lecithin–chitosan membranes showed no obvious creaming or phase separation. These results might be associated with the high surface charges which cause large repulsive force between the emulsion droplets.

The mean particle diameter of enzymatically crosslinked emulsions remained also unchanged (~370 nm) over a period of 9 days (**Figure 4**). Laccase-treated emulsions have also been shown to improve the emulsion stability against creaming, pH-influences, and salt (*Zeeb et al.*, 2011; *Zeeb et al.*, 2012). Interestingly, a color change of the enzyme-treated emulsion samples regardless of primary emulsifier type could be observed, whereas all emulsions were completely bleached after 20 days of storage (**Figure 5**). Potentially, encapsulated lutein was oxidized by the laccase leading to a decolorization of the oil phase. Indeed, when we further analyzed the unexpected color change of enzyme-treated emulsions using an instrumental colorimeter, a decrease in *b*-values over time was measured (**Figure 6**). The decrease of +*b*-values indicates that formation of free radicals catalyzed by the enzyme leading to a color change of emulsions. Laccases are well known to act on chromophoric groups such as phenols, amines, carotenoids or polyphenols (*Rodríguez Couto et al.*, 2006). The formation of radicals may lead to a breakdown of the chromophoric configuration of lutein which results in a decolorization of the carrier system. *Rodriguez et al.* (1999) already described similar effects. In their study, the enzyme seems to decolorize 23 different industrial dyes. The degradation of a textile dye by laccases purified from different fungi such as *Trametes hirsuta* and *Sclerotium rolfsii* was demonstrated by *Campos et al.* (2001). It was suggested that a laccase promoted oxidation led to the formation of anthranilic acid as a final degradation product. Moreover, emulsions containing laccase which was heat-treated (30 min, 85 °C) before addition to the secondary emulsions did not show a decrease in *b*-values indicating that the enzyme was completely inactivated. A laccase-induced aggregation of primary WPI-stabilized emulsions was observed leading to fluctuations in *b*- and *L*-values. All color measurements were also carried out at pH 7 to avoid an enzyme-promoted aggregation of emulsions. Our results showed that pH did not influence the mechanism of enzymatic decolorization. *L*-values were not affected by the addition of laccase regardless of pH or membrane thickness (Data not shown).

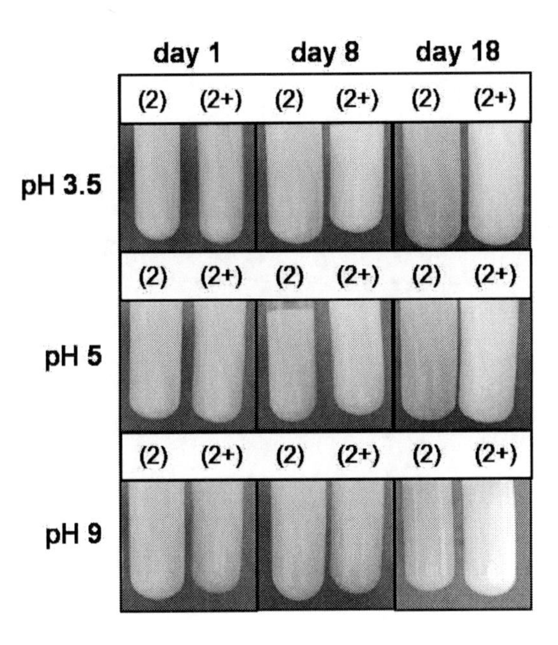


**Figure 5 Time-induced color changes of secondary (2) and laccase-treated secondary (2+) emulsions (5% w/w lutein-loading, 1% w/w Miglyol, 10 mM citrate buffer) containing droplets stabilized by WPI-beet pectin membranes over storage time (18 days) as a function of pH 3.5, 5 and 9.**

## Influence of membrane composition and treatment on lutein release in an acceptor oil medium

A modified membrane-free model was used to examine the lutein release of primary, secondary, and enzyme-treated secondary emulsions (5% (w/w) lutein-loading, 1% (w/v) oil, 10 mM citrate buffer, pH 3.5) into Miglyol (*Mitri et al.*, 2011). The release of encapsulated lutein was investigated over a period of 48 h and quantified by UV-vis-spectrophotometry. In **Figure 7**, the release profiles of primary, secondary and enzyme-treated secondary emulsions are shown. In general, we could observe biphasic release profiles. The initial release was followed by first-order kinetics involving lutein burst release, which might be induced by lutein directly adsorbed at the surface membrane pores (*Cho et al.*, 2003). The second phase followed a zero-order kinetic was observed for all samples, except the DTAB and WPI-stabilized primary emulsions (**Figure 7A** and **B**). Our results indicate that the layer composition, thickness, and treatment of the interfacial membrane influenced the release of the active ingredient into the surrounding medium. Primary DTAB-stabilized emulsions released 7.2% of encapsulated lutein in comparison to 0.24% of WPI-stabilized emulsions within the observation period of 48 h. This could be explained by the structure of the interfacial layer. WPI as a globular protein is known to form dense and compact interfacial membranes, which retard the encapsulated compound even better than DTAB and provides a layer that protrudes from the interface of an oil droplet to a depth of 1 to 2 nm (*Dalgleish et al.*, 1995). In contrary to WPI, surfactants such as DTAB are able to form interfacial membranes with a thickness of only a few angstrom (*Bos et al.*, 2001). In addition, the presence of DTAB micelles in the aqueous phase may have solubilized lutein from the interior of emulsion droplets, thus increasing its release rate.

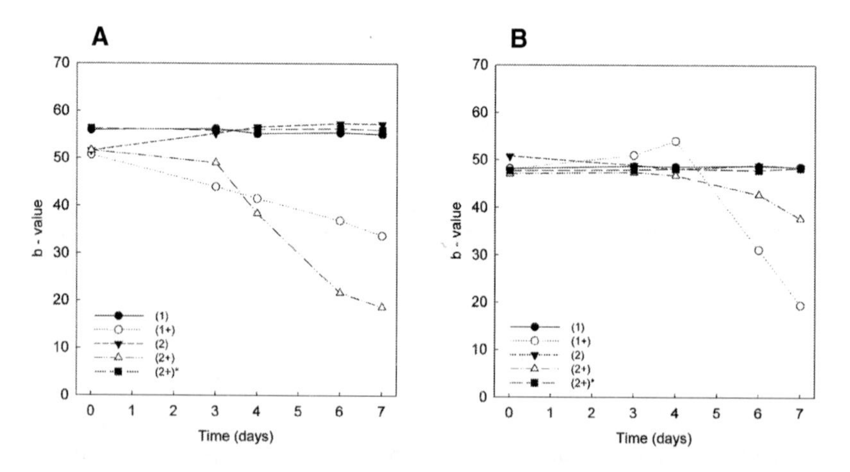


**Figure 6 pH-induced changes in tristimulus coordinate b (blue-yellow) of lutein-loaded WPI- (1) and WPI-beet pectin-stabilized (2) oil-in-water emulsions depending on heat (*)- and laccase (+)-treatment: pH 3.5 (A) and pH 7 (B).**

The adsorption of sugar beet pectin increased the layer thickness of primary emulsions containing oil droplets stabilized by either WPI or DTAB and thus influenced the release of

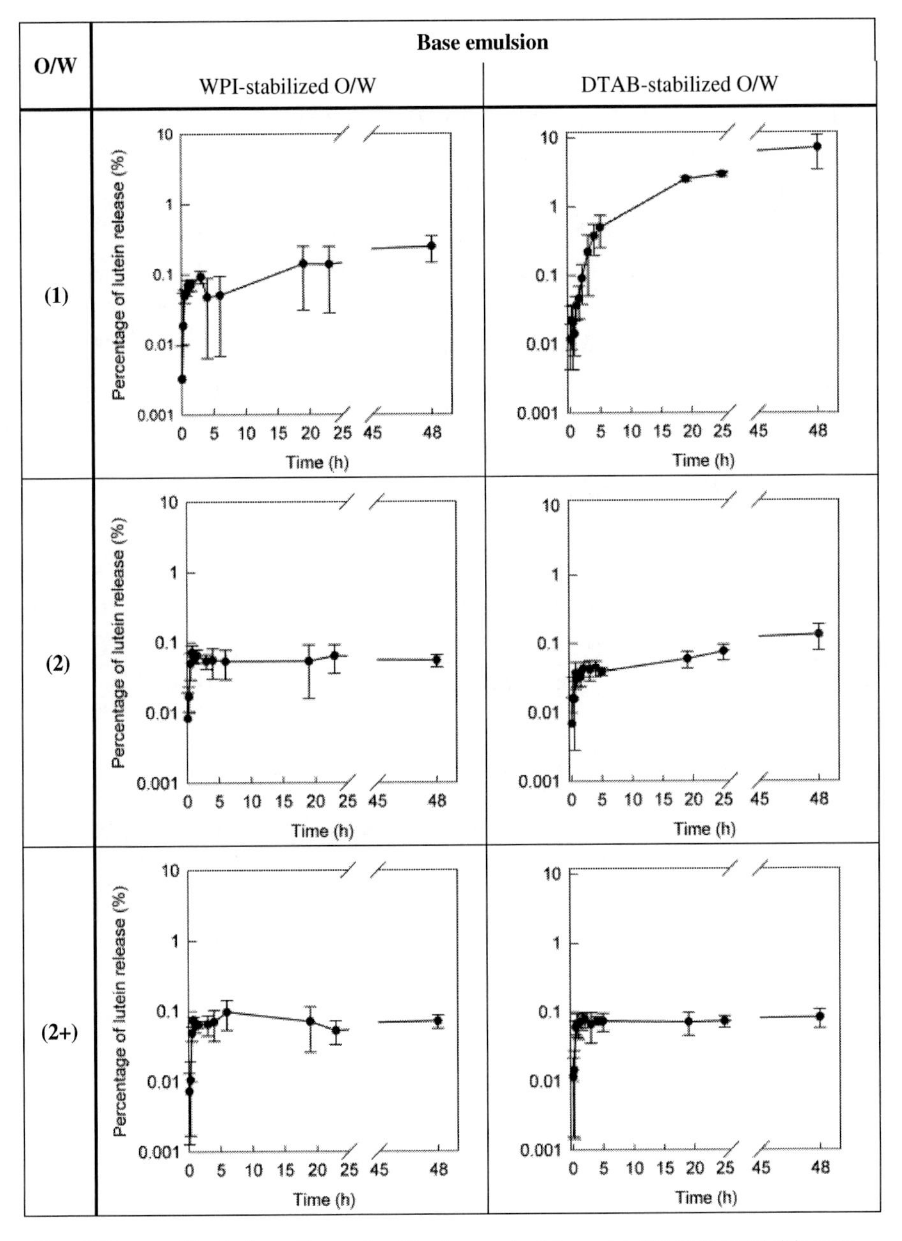


**Figure 7 Release profiles of primary (1), secondary (2) and enzyme treated secondary (2+) lutein-loaded oil-in-water emulsions (5% w/w lutein-loading, 1% w/v Miglyol, 10 mM citrate buffer, pH 3.5) over 48 h.**

lutein (**Figure 7C** and **D**). A significant decrease in the release of lutein from 0.24% to 0.054% ($p > 0.05$) was observed for secondary emulsions stabilized by WPI-beet pectin membranes (**Figure 7C**). Moreover, primary DTAB-stabilized emulsions released 7.2% of encapsulated lutein within the observation period, whereas beet pectin-DTAB coated emulsions released only 0.13% of lutein which was also described by *Benjamin et al* (2012). The release rates of hydrophobic volatile organic compounds were significantly higher for $\beta$-lactoglobulin-stabilized-emulsions compared to two layered $\beta$-lactoglobulin-pectin emulsions. A laccase-induced crosslinking of adsorbed sugar beet pectin did not significantly decrease the release of lutein compared to non-crosslinked secondary emulsions indicating that the formation of covalent crosslinks between single pectin molecules might not be sufficient enough to modify the porosity or permeability of the interfacial membrane.

**Influence of pH on lutein release of primary oil-in-water emulsions stabilized by WPI or DTAB**

A functional component such as lutein may be released at a particular site of action in response to an environmental trigger e.g. pH, salt or temperature (*McClements et al.*, 2009). Changes in environmental pH after deposition of biopolymers may lead to desorption of multilayers and therefore affect the release behavior of lutein. Therefore, primary, secondary, and laccase-treated secondary lutein-loaded emulsions were produced at pH 3.5 and subsequentially adjusted to pH 5 and 9 prior the release experiment.

**Figure 8** shows that the release of lutein was affected by changes in the pH of the surrounding medium. WPI-stabilized primary emulsions released significantly more lutein within the first hour at pH 5 in comparison to initial conditions ($p > 0.05$). After 48 h, 0.33% of lutein was released, whereas only 0.24% at a pH of 3.5 for WPI-stabilized emulsions. Interestingly, a decrease in lutein release was observed after 48 h when the pH of the surrounding phase was adjusted to pH 9. Our results showed that a release behavior of primary emulsions was influenced by the pH since environmental conditions mainly dominate the colloidal interactions between the droplets. An increase in release at pH 5.0 might be due to flocculation of the oil droplets stabilized by WPI. The primary emulsions had a positive charge (+28 mV) at pH 3.5, but the droplets became negatively charged (-48.2 ± 2.0 mV) when the pH was adjusted to pH 9. At a pH of 5.0, the droplets started to aggregate because this pH is close to the p*I* of the WPI and thus weak repulsive forces could not prevent droplet-droplet aggregation (> 6 μm) (*Demetriades et al.*, 1997). Although ζ-potential of WPI-stabilized emulsion decreased to a value of -48.2 ± 2.0 mV at pH 9 indicating strong repulsion between oil droplets an increase in particle size was observed (344.2 ± 6.2 nm) in comparison to pH-untreated emulsions (203 ± 4.0 nm). The droplets did not strongly remain aggregated at pH 5 after the pH was adjusted to 9 which indicated that the droplet flocculation at pH 5 was partially reversible within the experimental time scale (*Chanamai et al.*, 2002). Moreover, the decrease in lutein release at pH 9 might be due to conformation changes of the protein adsorbed at the interface which was previously described. *Monahan et al* (1995) showed an extensive irreversible unfolding of the protein under alkaline conditions (pH 9.0) due to disulfide-mediated polymerization reactions resulting in a denser layer.

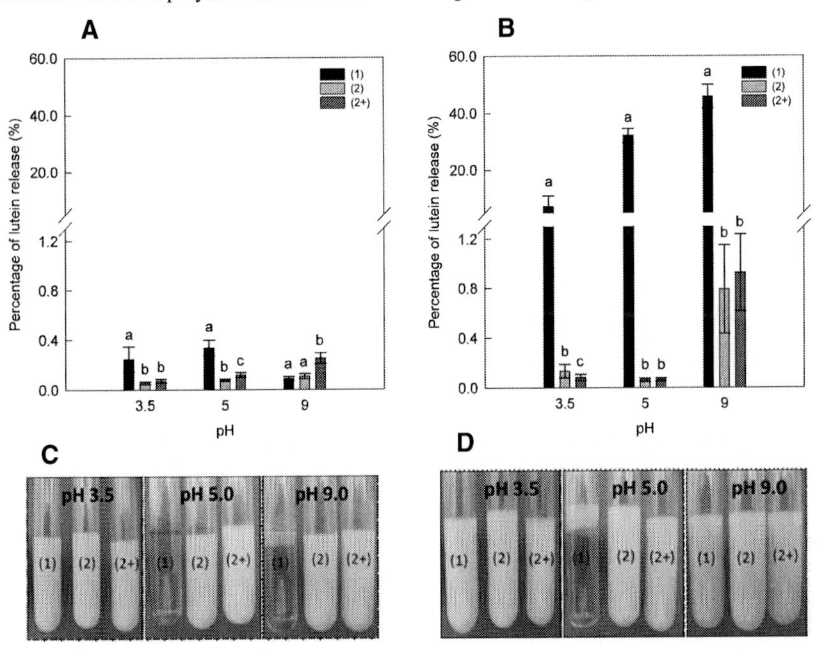


**Figure 8 pH-dependence of lutein release after 48 h from WPI-stabilized (A, C) and DTAB-stabilized (B, D) primary (1), secondary (2) and enzyme-treated secondary (2+) emulsion (5% w/w lutein-loading, 1% w/v oil).** [a, b, c] **indicate significant differences between samples ($p > 0.05$).**

DTAB-stabilized primary emulsions showed a significantly higher release at pH 5 and 9 compared to pH 3.5 ($p > 0.05$) (**Figure 8A**) which was attributed to a decrease in electrostatic interactions between the particles leading to a breakdown of the emulsion (*McClements*, 2004). When the pH was changed from acidic to alkaline conditions the ζ-potential decreased from +45.0 ± 5.7 to -20.0 ± 4.1, indicating that electrostatic repulsion could not prevent droplet-droplet encounters leading to coalescences and phase separation of the emulsion (**Figure 8**).

**pH-dependent release of lutein – Influence of layer thickness and treatment**

The optimum preparation conditions for the electrostatic deposition method required preparing stable oil-in-water emulsions without occurring flocculation or coalescence was previously carried out. Laccase was added to the secondary emulsions to promote interfacial crosslinking of adsorbed pectin molecules (*Zeeb et al.*, 2011; *Zeeb et al.*, 2012; *Zeeb et al.*, 2012).

The release of lutein from the interior of the droplets into the acceptor medium depending on pH and layer thickness is shown in **Figure 8**. Our results indicate that release could be retarded if thick interfacial membranes were built up. The release rates were lower for secondary and crosslinked secondary emulsions than in emulsions where droplets had been stabilized only by WPI or DTAB. The effect of pH on the release of lutein is also shown in **Figure 8**. The rates of lutein release showed a pH-dependent behavior which was attributed to the physical stability of the emulsions. Multilayered emulsions have previously shown to have a better stability against environmental stresses (*Moreau et al.*, 2003; *Gu et al.*, 2004; *Aoki et al.*, 2005). Furthermore, our study has been shown that laccase-induced crosslinking of adsorbed pectin in the interfacial membranes surrounding droplets in multilayered emulsions provides some limited improvement in terms of controlled release, in particular under alkaline conditions. In general, it is considered that crosslinking of adsorbed sugar beet pectin might alter the interactions between water and/or proteins and therefore change diffusion kinetics of lutein through the interfacial complex.

**Key insights**

In summary, we can highlight some key insights obtained from our study:

- WPI and DTAB may be used as emulsifiers to form stable lutein-loaded primary oil-in-water emulsions with different interfacial properties at acidic conditions. In contrast, it was not possible to form stable FG-stabilized emulsions loaded with lutein.

- The layer-by-layer electrostatic deposition method may be used to form a second layer on lutein-loaded oil-in-water emulsions by using sugar beet pectin. Laccase may induce crosslinking of adsorbed sugar beet pectin layer, but also may decolorize entrapped lutein.

- WPI may form a protective membrane which lower the release of incorporated lutein compared to DTAB. However, the presence of DTAB micelles may lead to an increased solubilization of lutein and therefore increase its release. In general, thicker double-layered interfacial membranes showed further reduced lutein release over time compared to single layered emulsions. Laccase induced covalent crosslinks between adsorbed pectin molecules might not be sufficient enough to modify the porosity or permeability of the interfacial membrane.

- Alterations of the pH may act as a release trigger due to changes in the physical stability of the emulsions. Multilayered and enzyme-treated emulsions showed enhanced stability against pH adjustments and may therefore be used for controlled release systems.

- The formation of free radicals induced by the addition of laccase promotes a color change of lutein-loaded emulsions. A subsequent heating step (> 85 °C) of laccase-treated emulsions could inactivate the enzyme and therefore prevent an enzymatic-induced colour change of beet pectin membranes.

## CONCLUSIONS

This study demonstrated that multilayering of emulsions using layer-by-layer depositioning can be a useful tool to create controlled release delivery systems in food. A laccase-induced crosslinking of adsorbed sugar beet pectin did not decrease the release of lutein in comparison to non-crosslinked secondary emulsions indicating that the formation of covalent crosslinks between single pectin molecules might not be sufficient enough to modify the porosity or permeability of the interfacial membrane. The formation of a tighter network structure may potentially change the diffusion properties of the interface and thus alter the release behavior of any encapsulated bioactive. Therefore, enzymatically designed biopolymers may be a potent building block to assemble distinct structures such as delivery and encapsulations systems. In addition, this study has been shown that enzyme technology might be a powerful tool to modify the color and appearance of emulsion based systems.

## ACKNOWLEDGEMENTS

We would like to thank Rudolf Wild GmbH & Co. KG (Heidelberg/Eppelheim, Germany), Arla Foods Ingredients (Viby, Denmark) and Herbstreith & Fox KG (Neuenbürg, Germany) for generously providing us with samples.

## REFERENCES

Ahmed, K., Li, Y., et al. (2012): Nanoemulsion- and emulsion-based delivery systems for curcumin: Encapsulation and release properties. Food Chemistry 132(2): 799-807.

Aoki, T., Decker, E. A., et al. (2005): Influence of environmental stresses on stability of O/W emulsions containing droplets stabilized by multilayered membranes produced by a layer-by-layer electrostatic deposition technique. Food Hydrocolloids 19(2): 209-220.

Barker Ii, F. M. (2010): Dietary supplementation: Effects on visual performance and occurrence of AMD and cataracts. Current Medical Research and Opinion 26(8): 2011-2023.

Benjamin, O., Silcock, P., et al. (2012): Multilayer emulsions as delivery systems for controlled release of volatile compounds using pH and salt triggers. Food Hydrocolloids 27(1): 109-118.

Bos, M. A., Van Vliet, T. (2001): Interfacial rheological properties of adsorbed protein layers and surfactants: A review. Advances in Colloid and Interface Science 91(3): 437-471.

Campos, R., Kandelbauer, A., et al. (2001): Indigo degradation with purified laccases from *Trametes hirsuta* and *Sclerotium rolfsii*. Biotechnology in the Textile Industry - Perspectives for the New Millennium 89: 131-139.

Chanamai, R., McClements, D. J. (2002): Comparison of gum arabic, modified starch, and whey protein isolate as emulsifiers: Influence of pH, $CaCl_2$ and temperature. Journal of Food Science 67(1): 120-125.

Chen, L., Remondetto, G. E., et al. (2006): Food protein-based materials as nutraceutical delivery systems. Trends in Food Science & Technology 17(5): 272-283.

Cho, Y. H., Shim, H. K., et al. (2003): Encapsulation of fish oil by an enzymatic gelation process using transglutaminase cross-linked proteins. Journal of Food Science 68(9): 2717-2723.

Dalgleish, D. G., Hallett, F. R. (1995): Dynamic light scattering: Applications to food systems. Food Research International 28(3): 181-193.

Dalgleish, D. G., Srinivasan, M., et al. (1995): Surface properties of oil-in-water emulsion droplets containing casein and Tween 60. Journal of Agricultural and Food Chemistry 43(9): 2351-2355.

Demetriades, K., Coupland, J. N., et al. (1997): Physical properties of whey protein stabilized emulsions as related to pH and NaCl. Journal of Food Science 62(2): 342-347.

Djordjevic, D., Kim, H. J., et al. (2004): Physical stability of whey protein-stabilized oil-in-water emulsions at pH 3: Potential ω-3 fatty acid delivery systems (Part A). Journal of Food Science 69(5): C351-C355.

Georgieva, D., Schmitt, V. r., et al. (2009): On the possible role of surface elasticity in emulsion stability. Langmuir 25(10): 5565-5573.

Gu, Y. S., Decker, A. E., et al. (2005): Production and characterization of oil-in-water emulsions containing droplets stabilized by multilayer membranes consisting of β-lactoglobulin, ι-carrageenan and gelatin. Langmuir 21(13): 5752-5760.

Gu, Y. S., Decker, E. A., et al. (2004): Influence of pH and ι-carrageenan concentration on physicochemical properties and stability of β-lactoglobulin-stabilized oil-in-water emulsions. Journal of Agricultural and Food Chemistry 52(11): 3626-3632.

Guzey, D., Kim, H. J., et al. (2004): Factors influencing the production of o/w emulsions stabilized by β-lactoglobulin-pectin membranes. Food Hydrocolloids 18(6): 967-975.

Guzey, D., McClements, D. J. (2006): Formation, stability and properties of multilayer emulsions for application in the food industry. Advances in Colloid and Interface Science 128-130: 227-248.

Jung, J., Wicker, L. (2012): Laccase mediated conjugation of sugar beet pectin and the effect on emulsion stability. Food Hydrocolloids 28(1): 168-173.

Kline, M. A., Duncan, S. E., et al. (2011): Light Wavelength Effects on a Lutein-Fortified Model Colloidal Beverage. Journal of Agricultural and Food Chemistry 59(13): 7203-7210.

Klinkesorn, U., Sophanodora, P., et al. (2005): Encapsulation of emulsified tuna oil in two-layered interfacial membranes prepared using electrostatic layer-by-layer deposition. Food Hydrocolloids 19(6): 1044-1053.

Kosaraju, S. L. (2005): Colon targeted delivery systems: Review of polysaccharides for encapsulation and delivery. Critical Reviews in Food Science and Nutrition 45(4): 251-258.

Li, Y., Hu, M., et al. (2010): Controlling the functional performance of emulsion-based delivery systems using multi-component biopolymer coatings. European Journal of Pharmaceutics and Biopharmaceutics 76(1): 38-47.

Like, M., Jia, Y., et al. (2010): Effects of homogenization models and emulsifiers on the physicochemical properties of β-carotene nanoemulsions. Journal of Dispersion Science & Technology 31: 986-993.

Littoz, F., McClements, D. J. (2008): Bio-mimetic approach to improving emulsion stability: Cross-linking adsorbed beet pectin layers using laccase. Food Hydrocolloids 22(7): 1203-1211.

Madene, A., Jacquot, M., et al. (2006): Flavour encapsulation and controlled release - A review. International Journal of Food Science and Technology 41(1): 1-21.

Malaki Nik, A., Wright, A. J., et al. (2010): Interfacial design of protein-stabilized emulsions for optimal delivery of nutrients. Food & Function 1(2): 141-8.

Matalanis, A., Jones, O. G., et al. (2011): Structured biopolymer-based delivery systems for encapsulation, protection, and release of lipophilic compounds. Food Hydrocolloids 25(8): 1865-1880.

McClements, D. J. (2002): Theoretical prediction of emulsion color. Advances in Colloid and Interface Science 97(1 - 3): 63-89.

McClements, D. J. (2004): Food emulsions: Principles, practice, and techniques. Boca Raton, CRC Press.

McClements, D. J., Decker, E. A., et al. (2009): Structural design principles for delivery of bioactive components in nutraceuticals and functional foods. Critical Reviews in Food Science and Nutrition 49(6): 577 - 606.

McClements, D. J., Decker, E. A., et al. (2007): Emulsion-based delivery systems for lipophilic bioactive components. Journal of Food Science 72(8): R109-R124.

McClements, D. J., Li, Y. (2010): Structured emulsion-based delivery systems: Controlling the digestion and release of lipophilic food components. Advances in Colloid and Interface Science 159(2): 213-228.

Mitri, K., Shegokar, R., et al. (2011): Lipid nanocarriers for dermal delivery of lutein: Preparation, characterization, stability and performance. International Journal of Pharmaceutics 414(1 - 2): 267-275.

Monahan, F. J., German, J. B., et al. (1995): Effect of pH and Temperature on Protein Unfolding and Thiol/Disulfide Interchange Reactions during Heat-Induced Gelation of Whey Proteins. Journal of Agricultural and Food Chemistry 43(1): 46-52.

Moreau, L., Kim, H. J., et al. (2003): Production and characterization of oil-in-water emulsions containing droplets stabilized by β-lactoglobulin-pectin membranes. Journal of Agricultural and Food Chemistry 51(22): 6612-6617.

Nik, A. M., Langmaid, S., et al. (2012): Digestibility and β-carotene release from lipid nanodispersions depend on dispersed phase crystallinity and interfacial properties. Food and Function 3(3): 234-245.

Rodríguez Couto, S., Toca Herrera, J. L. (2006): Industrial and biotechnological applications of laccases: A review. Biotechnology Advances 24(5): 500-513.

Rodríguez, E., Pickard, M. A., et al. (1999): Industrial dye decolorization by laccases from ligninolytic fungi. Current Microbiology 38(1): 27-32.

Sandra, S., Decker, E. A., et al. (2008): Effect of interfacial protein cross-linking on the in vitro digestibility of emulsified corn oil by pancreatic lipase. Journal of Agricultural and Food Chemistry 56(16): 7488-7494.

Shefer, A., Shefer, S. (2003): Novel encapsulation system provides controlled release of ingredients. Food Technology 57(11): 40-42.

Sowbhagya, H. B., Sampathu, S. R., et al. (2004): Natural colorant from marigold-chemistry and technology. Food Reviews International 20(1): 33-50.

Surh, J., Decker, E. A., et al. (2006): Properties and stability of oil-in-water emulsions stabilized by fish gelatin. Food Hydrocolloids 20(5): 596-606.

Vishwanathan, R., Wilson, T. A., et al. (2009): Bioavailability of a nanoemulsion of lutein is greater than a lutein supplement. Nano Biomedicine and Engineering 1(1): 38-49.

Wilde, P., Mackie, A., et al. (2004): Proteins and emulsifiers at liquid interfaces. Advances in Colloid and Interface Science 108-109: 63-71.

Wooster, T. J., Golding, M., et al. (2008): Impact of oil type on nanoemulsion formation and Ostwald ripening Stability. Langmuir 24(22): 12758-12765.

Yang, X., Tian, H., et al. (2012): Stability of citral in emulsions coated with cationic biopolymer layers. Journal of Agricultural and Food Chemistry 60(1): 402-409.

Zaidel, D. N. A., Chronakis, I. S., et al. (2013): Stabilization of oil-in-water emulsions by enzyme catalyzed oxidative gelation of sugar beet pectin. Food Hydrocolloids 30(1): 19-25.

Zeeb, B., Fischer, L., et al. (2011): Cross-linking of interfacial layers affects the salt and temperature stability of multilayered emulsions consisting of fish gelatin and sugar beet pectin. Journal of Agricultural and Food Chemistry 59(19): 10546-10555.

Zeeb, B., Gibis, M., et al. (2012): Crosslinking of interfacial layers in multilayered oil-in-water emulsions using laccase: Characterization and pH-stability. Food Hydrocolloids 27(1): 126-136.

Zeeb, B., Gibis, M., et al. (2012): Influence of interfacial properties on Ostwald ripening in crosslinked multilayered oil-in-water emulsions. Journal of Colloid and Interface Science 387(1): 65-73.

# CHAPTER 6

## Heat- and laccase-induced crosslinking of adsorbed biopolymer nanoparticles affects pH-stability of WPI-stabilized oil-in-water emulsions

*Benjamin Zeeb*[1], *Hanna Salminen*[1], *Lutz Fischer*[2], *Jochen Weiss*[1]

[1] Department of Food Physics and Meat Science, University of Hohenheim, Garbenstrasse 21/25, 70599 Stuttgart, Germany

[2] Department of Food Biotechnology, University of Hohenheim, Garbenstrasse 25, 70599 Stuttgart, Germany

Reprinted from "*Impact of heat and laccase on the pH and freeze-thaw stability of oil-in-water emulsions stabilized by adsorbed biopolymer nanoparticles*", Zeeb, B., Salminen, H., Fischer, L., Weiss, J., Food Biophysics, 2014, 9(2), p. 125-137 with permission from Springer.

## ABSTRACT

The enzymatic cross-linking of adsorbed biopolymer nanoparticles formed between whey protein isolate (WPI) and sugar beet pectin using the complex coacervation method was investigated. A sequential electrostatic depositioning process was used to prepare emulsions containing oil droplets stabilized by WPI – nanoparticle – membranes. Firstly, a finely dispersed primary emulsion (10% w/w Miglyol oil, 1% w/w WPI, 10 mM acetate buffer at pH 4) was produced using a high-pressure homogenizer. Secondly, a series of biopolymer particles were formed by mixing WPI (0.5% w/w) and pectin (0.25% w/w) solutions with subsequent heating above the thermal denaturation temperature (85°C, 20 min) to prepare dispersions containing particles in the submicron range. Thirdly, nanoparticle-covered emulsions were formed by diluting the primary emulsion into coacervate solutions (0-0.675% w/w) to coat the droplets. Oil droplets of stable emulsions with different interfacial membrane compositions were subjected to enzymatic cross-linking. We used cross-linked multilayered emulsions as a comparison. The pH stability of primary emulsions, biopolymer complexes and nanoparticle-coated base emulsions, as well as multilayered emulsions, was determined before and after enzyme addition. Moreover, freeze-thaw stability (-9°C for 22 h, 25°C for 2 h) of nanoparticle-coated emulsions was not affected by laccase. Results indicated that cross-linking occurred exclusively in the multilamellar layers and not between adsorbed biopolymer nanoparticles. Results suggest that the accessibility of distinct structures may play a key role for biopolymer-cross-linking enzymes.


**Keywords:** Multilayered emulsions; Coacervates; Layer-by-layer depositioning, Sugar beet pectin; Whey protein isolate; Laccase; Stability

## INTRODUCTION

Food emulsions are commonly stabilized with a layer of biopolymers (proteins or polysaccharides) or food-grade surfactants (*Leroux et al.*, 2003; *McClements*, 2004; *Chee et al.*, 2008; *Iwanaga et al.*, 2008). Depending on the emulsifier type, they can provide sufficiently strong repulsive forces between the emulsion droplets, either through electrostatic or steric interactions. However, electrostatically stabilized oil-in-water emulsions are sensitive to pH, salt, heating, freezing, chilling, and dehydration (*Dickinson*, 1992; *Demetriades et al.*, 1997; *Chanamai et al.*, 2002; *Dickinson*, 2010). Approaches to create novel emulsion structures are emerging to overcome these instabilities. The formation of multilayered interfaces using a layer-by-layer (LbL) electrostatic deposition technique utilizes the electrostatic attraction of oppositely charged electrolytes and assembles them into two, three or multiple layers around the oil droplets (*Guzey et al.*, 2006). As the thickness of the interfacial layers increases, the oil-in-water emulsions have shown increased stability against various environmental stresses A variety of electrolytes can be used in the LbL technique, including surfactants, biopolymers, biopolymer complexes, and colloidal particles (*Guzey et al.*, 2006; *Guzey et al.*, 2006; *Gu et al.*, 2007).

Biopolymer complexes or coacervates can be assembled by phase separating mixed biopolymer systems (*Turgeon et al.*, 2007). Proteins have a net positive charge below their isoelectric point (p*I*) and a net negative charge above their p*I*, whereas anionic polysaccharides have a negative charge above their p*I*. Complex coacervation occurs through associative separation, where oppositely charged protein and polysaccharide molecules form electrostatic complexes at a pH just below the p*I* of the proteins (*Turgeon et al.*, 2007). The formation and stability of biopolymer complexes or coacervates depends on the biopolymer ratio, pH, ionic strength, temperature, charge density, and co-solvents (*Turgeon et al.*, 2007; *Jones et al.*, 2008; *Chanasattru et al.*, 2009; *Jones et al.*, 2009; *Gentes et al.*, 2010; *Jones et al.*, 2010; *Jones et al.*, 2010). Our previous study revealed that WPI-pectin coacervates can be electrostatically adsorbed to protein-stabilized emulsion interfaces (*Salminen et al.*, 2013). Furthermore, the stability of emulsions coated with a coacervate layer improved their stability.

Recently, enzymatic crosslinking of multilayered interfaces has shown further enhancements in their physical stability. Enzymes such as laccase (*Littoz et al.*, 2008; *Zeeb et al.*, 2011; *Chen et al.*, 2012; *Zeeb et al.*, 2012), horseradish peroxidase (*Rauf et al.*, 2006; *Kim et al.*,

2010), or lactate oxidase (*Kim et al.*, 2010) have been reported to induce crosslinks in multilayered structures. We reported in our previous studies that laccase cross-linked the beet pectin layer in a fish gelatin-beet pectin multilayered oil-in-water emulsion, thereby improving its salt, pH, heat, and freeze-thaw stability (*Zeeb et al.*, 2011; *Zeeb et al.*, 2012). In a study by Chen et al. (*Chen et al.*, 2012), crosslinking of WPI-beet pectin complex coacervates led to formation of more rigid and more elastic gels with highly organized microstructures than the gels without crosslinking. Therefore, enzymatic cross-linking offers an opportunity to engineer and produce biopolymer particles with different particle sizes and properties to create novel structures and delivery systems for the encapsulation, protection and delivery of bioactive or functional ingredients.

The objectives of this study were to compare the influence of laccase on the stability of multilayered emulsions and emulsions covered with coacervates. Using two structurally different, but otherwise the same, interfacial layering (multiple layers *vs.* nanoparticles) biopolymers, we can assess their effect on emulsion stability more precisely. Our hypothesis was that laccase may promote covalent cross-links between beet pectin molecules, WPI molecules and/or WPI and beet pectin molecules via ferulic acid and/or tyrosine bonds. On the basis of previous studies (*Zeeb et al.*, 2011; *Zeeb et al.*, 2012), we expected that cross-linking would increase the stability of oil-in-water emulsions containing WPI-beet pectin layers (i.e. multilayered emulsion). Furthermore, we hypothesized that cross-linking of coacervates and their subsequent depositioning on the emulsion droplet interface increases their stability. For this purpose, we firstly prepared WPI-beet pectin coacervates with or without heat treatment, which were subsequently treated with laccase. Our hypothesis was that the type of heat treatment may affect the stability of the protein-polysaccharide complexes in the absence or presence of laccase. The effect of heating was demonstrated in a study by Jones et al. (*Jones et al.*, 2010), in which β-lactogloblin-pectin complexes showed increased pH stability upon heating above the thermal denaturation temperature of the proteins. This is because heating leads to partial or complete unfolding of proteins, thus exposing reactive groups from the hydrophobic core.

## MATERIALS AND METHODS

**Materials.** WPI (#DSE 9273) was purchased from Fonterra GmbH (Hamburg, Germany). WPI is composed of a mixture of β-lactoglobulin, α-lactalbumin, bovine serum albumin, and other proteins, such as caseins. The p*I* of β-lactoglobulin is 5.3, whereas the p*I* of α-lactalbumin is 4.1, and thus the WPI has a p*I* close to 5 (*Dickinson*, 2010). Sugar beet pectin (#1 09 03 135) was donated by Herbstreith & Fox KG (Neuenbürg, Germany) and used without further purification. The degree of esterification of the beet pectin was 55%, as stated by the manufacturer. Laccase (#0001437590, from *Trametes versicolor*) was obtained from Sigma–Aldrich Co. (Steinheim, Germany). The Laccase obtained was reported to have 13.6 units per mg (AU) of enzyme. Mygliol 812N, a medium chain triacylglyceride mixture, was purchased from Sassol Germany GmbH (Brunsbüttel, Germany). Trans-ferulic acid (#46278, purity ≥ 99.0%) was obtained from Sigma-Aldrich Co. (Steinheim, Germany). L-tyrosine (#K40556871 033, purity ≥ 99.0%) was purchased from Merck KGaA (Darmstadt, Germany). L-methionine (#90M006934V, purity ≥ 98.0%) was obtained from Sigma-Aldrich Co. (Steinheim, Germany). Acetic acid glacial (#3738.5, purity = 100%), sodium acetate (#6773.3, purity ≥ 99.0%) and sodium azide (#K305.1, purity ≥ 99.0%) were obtained from Carl Roth GmbH & Co. KG (Karlsruhe, Germany). Analytical grade hydrochloric acid (HCl) and sodium hydroxide (NaOH) were also purchased from Carl Roth GmbH & Co. KG (Karlsruhe, Germany). Distilled water was used for the preparation of all samples.

### A. Preparation of enzyme-treated multilayered emulsions.

**Solution preparation.** Aqueous emulsifier solutions were prepared by dispersing 1% w/w WPI powder into 10 mM acetate buffer (pH 4). Sugar beet pectin solutions were prepared by dispersing 1% w/w powdered pectin into 10 mM acetate buffer at pH 4, followed by stirring overnight to ensure complete hydration. The pH was then adjusted to 4 using 1 M HCl and/or 1 M NaOH. Sodium azide (0.02% w/w) was used as an antimicrobial agent. Enzyme solutions were prepared by dispersing enzyme powder into 10 mM acetate buffer (pH 4) followed by stirring for 30 min.

**Preparation of emulsions.** Primary base emulsions were prepared by homogenizing 10% w/w Miglyol and 90% w/w aqueous WPI solution at room temperature. A coarse preemulsion was formed by blending oil and emulsifier solution in a high shear blender (Standard Unit, IKA Werk GmbH, Germany) for 2 min and premixes were then passed through a high-pressure homogenizer (Avestin, Inc., Ottawa, Ontario, Canada) three times at 1000 bar.

Primary emulsions were diluted with acetate buffer (10 mM, pH 4) to obtain a final oil droplet concentration of 0.5% w/w.

Multilayered emulsions (i.e. secondary emulsions) were prepared by diluting the primary emulsions with 10 mM acetate buffer (pH 4) and aqueous beet pectin solution (pH 4) (1:20) to produce secondary emulsions with a 0.5% w/w oil, 0.045% w/w WPI and 0.1% w/w beet pectin concentration.

The enzyme-treated multilayered emulsions were prepared by adding laccase to the secondary emulsion (0.5% w/w oil, 0.045% w/w WPI, 0.1% w/w beet pectin) by using a vortex. An enzyme/polysaccharide ratio of 0.24/4 mg, respectively, equivalent to 5 AU, was sufficient to promote cross-linking of biopolymers, as previously described by *Zeeb et al.* (2012).

**B. Formation of biopolymer nanoparticles using complex coacervation.**

**Solution preparation.** Powdered WPI and pectin were dissolved in 10 mM sodium acetate (pH 7.0) containing sodium azide (0.02% (w/w)) as an antimicrobial agent, and stirred at ambient temperature for at least 4-8 hours. After the protein and pectin were found to reduce the pH after solubilization, the solutions were adjusted back to pH 7.0 using 1.0 and 0.1 N sodium hydroxide solutions before being mixed together. After mixing, the final protein concentration in the solution was 0.5% (w/w), while the pectin concentration was 0.25% (w/w).

**Preparation of nanoparticles.** Various heat treatments were performed before and after complex coacervation procedure to produce dispersions containing particles in the submicron range.

i. Native coacervates: The mixed WPI-pectin solution was adjusted to pH 4.0 and stirred for 30 min without heat treatment.

ii. Heated coacervates: The mixed WPI-pectin solution was adjusted to pH 4.0 and stirred for 30 min. Then the mixed solution was heat treated at 85°C for 20 min and cooled at room temperature for 2 h.

iii. Denaturated WPI-pectin complexes: Only the WPI solution (pH 7) was heated at 85°C for 20 min, cooled at room temperature, then mixed together with pectin (pH 7), and adjusted to pH 4.

**Enzyme-treatment of nanoparticles.** Laccase was added to all protein-polysaccharide complexes to induce protein-protein, polysaccharide-polysaccharide, and/or protein-polysaccharide crosslinks, respectively. We used the same enzyme/polysaccharide ratio as described above to promote crosslinking.

**C. Formation of crosslinked nanoparticle-coated base emulsions.**

**Coating and crosslinking of base emulsions.** Coacervate-coated emulsions were prepared by mixing the primary base emulsions with coacervates and 10 mM acetate buffer (pH 4) using a high shear blender (Standard Unit, IKA Werk GmbH, Germany) for 2 min to produce a series of emulsions (0.5% (w/w) oil droplet concentration) with different coacervate concentrations (0 – 0.675% (w/w)). Again, an enzyme/polysaccharide ratio of 0.24 mg/4 mg was used to promote crosslinking of biopolymers.

**Particle size determination.** Particle size distributions were determined using both static and dynamic light scattering techniques due to the wide range of droplet sizes the emulsions and coacervates had. Static light scattering was performed using a static light scattering instrument (Horiba LA-950, Retsch Technology GmbH, Haan, Germany). Samples were withdrawn and diluted to a droplet concentration of approximately 0.005% (w/w) with an appropriate buffer to prevent multiple scattering effects. The instrument measured the angular dependence of the intensity of the laser beam scattered by the dilute emulsions and then used the Mie theory to calculate the droplet size distributions that gave the best fit between theoretical predictions and empirical measurements. A refractive index ratio of 1.08 (ratio of the indices between the oil and water phase) was used. The method is able to accurately measure the diameter of droplets in the micrometer range.

Dynamic light scattering was used to determine the particle diameters of dispersions in the submicron range (Nano ZS, Malvern Instruments, Malvern, UK). Samples were diluted to a droplet concentration of approximately 0.005% (w/w) with an appropriate buffer to prevent multiple scattering effects. The foundation of this technique is based on the scattering of light by moving particles due to Brownian motion in a liquid (*Dalgleish et al.*, 1995). The movement of the particles is then related to the size of the particles. The instrument reports the mean particle diameter (z-average) and the polydispersity index (PDI) ranging from 0 (monodisperse) to 0.50 (very broad distribution).

**$\zeta$-Potential measurements.** Emulsions and coacervate suspensions were diluted to a droplet concentration of approximately 0.005% (w/w) with an appropriate buffer. Diluted emulsions were then loaded into a cuvette of a particle electrophoresis instrument (Nano ZS, Malvern Instruments, Malvern, UK), and the ζ-potential was determined by measuring the direction and velocity that the droplets moved in the applied electric field. The ζ-potential measurements were reported as the average and standard deviation of measurements made from two freshly prepared samples, with 3 readings made per sample.

**Optical microscopy.** All emulsions samples were gently mixed before analysis using a vortex to ensure emulsion homogeneity. One drop of emulsions was placed on an objective slide and then covered with a cover slip. Light microscopy images were taken with an axial mounted Canon Powershot G10 digital camera (Canon, Tokyo, Japan) mounted on an Axio Scope optical microscope (A1, Carl Zeiss Microimaging GmbH, Göttingen, Germany).

**Isothermal titration calorimetry.** An isothermal titration calorimeter (ITC 2G Isothermal Titration Calorimeter, TA Instruments, Lindon, UT 84042, US) was used to measure the laccase-induced oxidation of 1 mM ferulic acid, 1 mM L-tyrosine, and 1 mM L-methionine at 22 °C. Twenty 10 μL aliquots of ferulic acid, L-tyrosine, and L-methionine were injected sequentially into a 996 μL titration cell containing laccase (1 mg/10 ml). Each injection lasted 200 s with an interval of 600 s between each injection. The solution in the titration cell was stirred at 200 revolutions per minute throughout the experiment. All solutions were tempered and degassed for 30 min. prior the measurement was carried out.

**Emulsion environmental stress tests.** We determined the influence of various kinds of environmental stresses on the mean particle diameter (z-average), ζ-potential and microstructure of the emulsion samples produced.

*pH stability.* All the different emulsion and coacervate samples were adjusted with a range of different levels of pH (3-8) using 0.1 and 1 M HCl and/or NaOH, respectively. All samples were kept for 1 min after reaching the pH value before transferring (10 ml) into glass test tubes and analyzing the mean particle distribution and ζ-potential after 24 h. Pictures of all samples were taken using a digital camera.

*Freeze-thaw cycling.* Emulsions (10 ml) were transferred into glass test tubes and were incubated in a -9°C salt-water bath for 22 h. After incubation, the emulsion samples were thawed by incubating in a 25°C water bath for 2 h. This freeze-thaw cycle was repeated twice and its influence on the mean particle diameter and microstructure was measured after each cycle.

**Statistical analysis.** All measurements were repeated at least 3 times using duplicate samples. Means and standard deviations were calculated from these measurements using Excel (Microsoft, Redmond, VA, USA).

## RESULTS AND DISCUSSION

### A. Characterization and pH-stability of primary emulsions

Firstly, we examined the mean particle size distribution and ζ-potential of freshly prepared primary base emulsions at pH 4. WPI is known to have surface active properties and can therefore act as an emulsifier in oil-in-water emulsions (*Dickinson*, 1992; *McClements*, 2004; *Dickinson*, 2010). Dynamic light-scattering measurements showed monomodally distributed primary emulsions with a mean particle diameter of 229 ± 3 nm (**Figure 1**). The polydispersity index was 0.208 ± 0.015 indicating a narrow distribution – a fact that is important in preparing multilayered emulsions. The electrical charge of the WPI-stabilized oil droplets was positive (+46 ± 1 mV) at pH 4 because of the high p*I* value of the protein (**Figure 1**). Furthermore, emulsions stabilized by a single WPI-coating were sTable over the entire measurement duration; the mean particle diameter and ζ-potential remained unchanged over a period of 7 d, which is in line with studies published previously (*Demetriades et al.*, 1997; *Chanamai et al.*, 2002). At pH 4, the physical stability of WPI-stabilized emulsions is due to the relatively strong electrostatic repulsion forces acting between the oil droplets (*Demetriades et al.*, 1997; *Dickinson*, 2010).

Secondly, we determined the pH stability (3 to 8) of the primary base emulsions. The primary emulsion had a positive charge (+69 ± 7 mV) at pH 3 (**Figure 1**). When the primary emulsion was adjusted to pH 8, the oil droplets became negatively charged (-56 ± 2 mV). Moreover, at pH 5, the droplets started to aggregate strongly because this pH was close to the p*I* of the WPI, and thus, repulsive forces were too inefficiently low to prevent droplet-droplet aggregation and collisions. Therefore, droplet flocculation could be observed at pH 5 (z-average particle diameter > 11 μm) (**Figure 1**).

### B. Multilayered emulsions

#### Influence of enzymatic cross-linking on pH-stability of multilayered emulsions

In our previous work, we already identified the optimum preparation conditions required to prepare stable oil-in-water emulsions containing droplets with multilaminar interfaces using electrostatic depositioning without incurring flocculation or coalescence of particles (*Zeeb et al.*, 2011; *Zeeb et al.*, 2012).

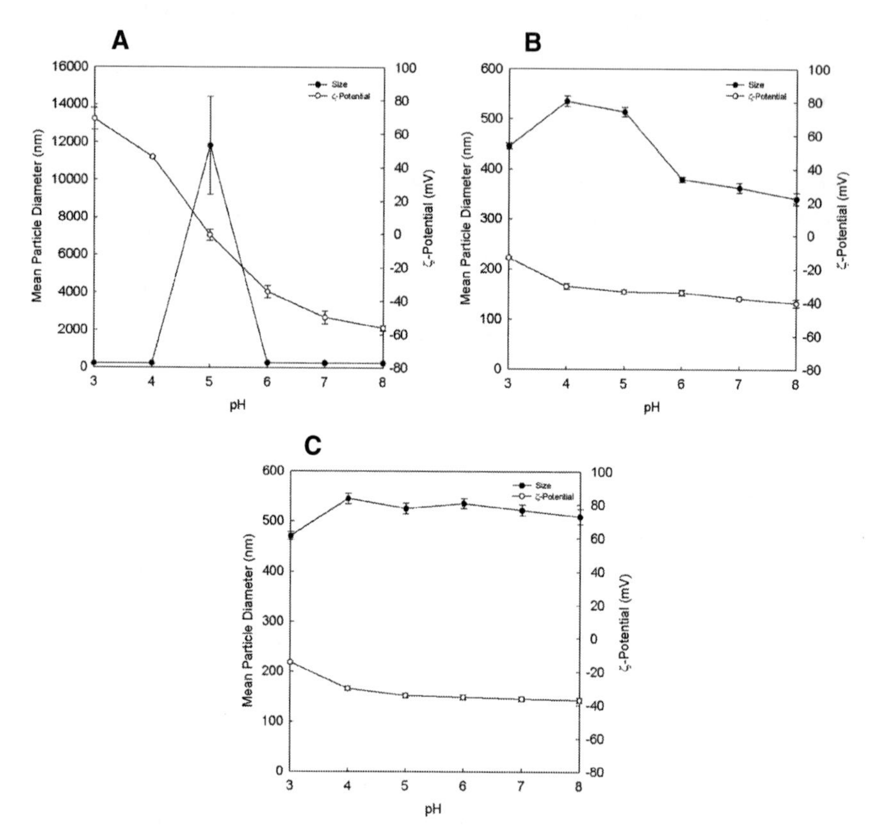


**Figure 1 Mean droplet diameter and ζ-potential of primary WPI-stabilized (A), secondary WPI-sugar beet pectin-stabilized (B), and laccase-treated secondary WPI-sugar beet pectin-stabilized emulsions (C) (0.5% w/w Miglyol) as a function of pH (3 to 8).**

The enzyme laccase was added to the multilayered emulsions to promote the interfacial cross-linking of adsorbed pectin molecules. The objective of these experiments was to determine the effects of different pH (3 to 8) on the mean particle diameter and ζ-potential of multilayered and enzyme-treated multilayered emulsions (**Figure 1**). The ζ-potential of the droplets in the multilayered emulsions changed form -13 ± 1 mV to -40 ± 2 mV as the pH was increased from 3 to 8 (**Figure 1B**). Therefore, at pH 8, the pectin did not remain attached to the surface of the primary emulsion droplets. This phenomenon could be attributed to the fact that the surface of the WPI-coated droplets at this pH is negatively charged, and thus, the electrostatic repulsion between the adsorbed WPI molecules and the pectin polymers cause them to detach from the droplet surface. The mean particle diameter of the multilayered emulsions depending on the pH showed a decrease in particle size when the pH was increased (**Figure 1B**). Finally, when the pH of laccase-treated multilayered emulsion was increased from pH 3 to 8, the droplets became increasingly negatively charged and the mean particle size remained constant (**Figure 1C**) suggesting that laccase had successfully linked the pectin molecules together preventing a detachment, which was confirmed in our previous study (*Zeeb et al.*, 2012).

### C. Formation of small-sized protein-polysaccharide nanoparticles

#### Electrical properties of WPI and sugar beet pectin solutions

Initially, the influence of pH (3 to 7) on the ζ-potential of WPI and sugar beet pectin solutions was measured (**Figure 2**). The ζ-potential of WPI solutions changed from positive (+18.7 ± 1.9 mV) to negative (-21.2 ± 1.6 mV) as the pH was increased from 3 to 7. The point of zero charge (p*I*) for WPI was around 4.6, which was in accordance with other studies (*Jones et al.*, 2008; *Chanasattru et al.*, 2009). Sugar beet pectin displayed the characteristic behaviour of an anionic polyelectrolyte. The ζ-potential of sugar beet pectin solutions remained negative as the pH was increased from 3 to 7 (**Figure 2**) which could be attributed to the fact that the $pK_a$ values of pectin is around pH 3.5 (*Guzey et al.*, 2006; *Iwanaga et al.*, 2008).

#### Effect of heat treatment in complex coacervations

The objective of these experiments was, firstly, to identify the effect of heat treatment on the particle size of the protein-polysaccharide complexes, which could then be further adsorbed onto the interfacial membranes of WPI-stabilized droplets by electrostatic depositioning. The following complex coacervations (at pH 4) were prepared: (**i**) native coacervates, (**ii**) denatured WPI-pectin complexes and (**iii**) heated coacervates. The effect of thermal treatment on the particle size distribution and polydispersity index of the WPI-beet pectin complexes was then investigated by dynamic light-scattering (**Table 1**).

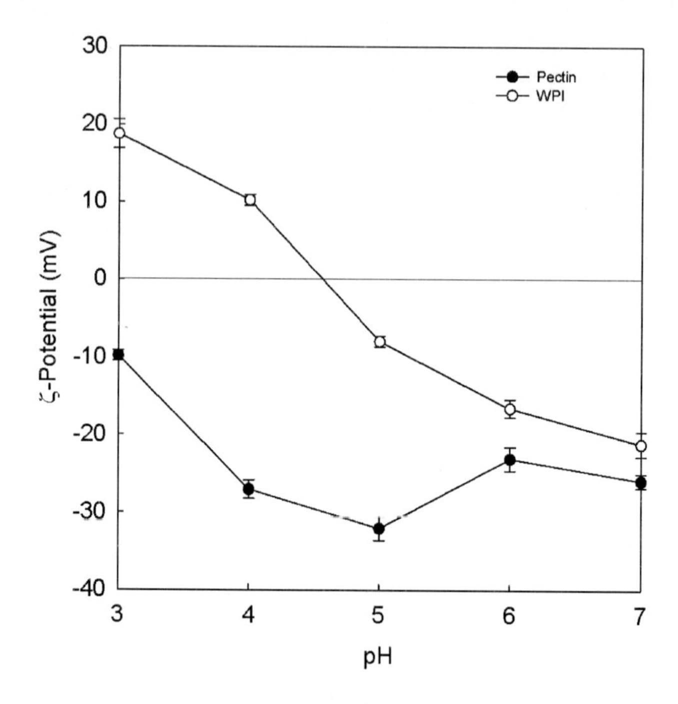


**Figure 2 pH-dependence of electrical properties (ζ-potential) of aqueous biopolymer solutions (10 mM acetate buffer) containing WPI (0.5% w/w) and sugar beet pectin (0.5% w/w).**

Our results indicated that the effect of thermal treatment influenced the formation of complex coacervations with particle sizes ranging in the submicron region. Native and heated coacervate solutions showed the smallest mean particle diameters with 195.6 ± 3.8 and 236.9 ± 3.4, respectively (**Table 1**). Moreover, the polydispersity indices confirmed a narrow distribution of these particles. *Jones et al.* (2009) also indicated an increase in particle diameter depending on holding temperature. The authors concluded that the initial increase in size may be attributed to an increasing incorporation of denaturated protein molecules into individual biopolymer particles followed by structural rearrangements. Furthermore, when we heated the WPI solution separately and mixed it together with pectin solution to prepare denaturated WPI-pectin complexes (pH 4), the particle diameter increased almost two-fold in size (407.4 ± 9.6) (**Table 1**).

**Table 1 Influence of heat treatment (85°C, 20 min) on mean particle diameter (z-average) and polydispersity index (PDI) of native coacervates, denaturated WPI-pectin complexes and heated coacervates (0.5% w/w WPI, 0.25% w/w sugar beet pectin, 10 mM acetate buffer).**

| | native coacervates | | denaturated WPI-pectin complex | | heated coacervates | |
|---|---|---|---|---|---|---|
| pH | Z-Ave (nm) | PDI (-) | Z-Ave (nm) | PDI (-) | Z-Ave (nm) | PDI (-) |
| 3 | 3113.8 ± 211.1 | 1.000 ± 0.000 | 1921.0 ± 646.9 | 0.570 ± 0.159 | 8525.6 ± 1083.5 | 0.379 ± 0.127 |
| 4 | 195.6 ± 3.8 | 0.187 ± 0.014 | 407.4 ± 9.6 | 0.543 ± 0.067 | 236.9 ± 3.4 | 0.238 ± 0.010 |
| 5 | 224.6 ± 15.6 | 0.603 ± 0.128 | 483.2 ± 25.4 | 0.535 ± 0.040 | 250.6 ± 4.5 | 0.220 ± 0.010 |
| 6 | 282.8 ± 18.1 | 0.990 ± 0.020 | 105.1 ± 27.4 | 0.604± 0.089 | 250.8 ± 6.8 | 0.215 ± 0.009 |
| 7 | 284.7 ± 34.9 | 0.884 ± 0.092 | 80.3 ± 4.4 | 0.705 ± 0.024 | 249.8 ± 5.1 | 0.199 ± 0.012 |
| 8 | 379.7 ± 45.0 | 0.786 ± 0.161 | 106.6 ± 45.0 | 0.620 ± 0.128 | 243.9 ± 5.4 | 0.199 ± 0.021 |

These results are in line with previously published studies (*Jones et al.*, 2009). It was found that the size and concentration of aggregates formed by heating globular protein solutions increased with increasing holding time (*Hoffmann et al.*, 1996; *Hoffmann et al.*, 1997). Therefore, *Jones et al.* (2009, 2010) suggested that globular proteins such as WPI in protein-polysaccharide mixtures tend to denaturate and aggregate by a fairly similar mechanism to globular proteins alone at higher temperatures. In general, the formation of stable biopolymer associations depends on many factors such as protein-polysaccharide ratio, ionic strength, pH, charge density, and temperature (*Turgeon et al.*, 2007; *Gentes et al.*, 2010; *Jones et al.*, 2010).

**pH-stability of coacervates**

The objective of these experiments was to investigate the pH stability of native, heated and denatured protein-polysaccharide complexes. We hypothesized that heat-treated complexes achieve a higher stability against pH-induced dissociation in comparison to non-heat-treated complexes, since proteins partially or completely unfold upon heating above their thermal denaturation temperature exposing some reactive groups. In particular, sugar beet pectins are reported to have a significantly higher proteinaceous moiety (10.4%) compared to apple (1.6%) or citrus (3-3.3%) pectins, which may also promote protein-polysaccharide cross-links upon heating (*Thibault*, 1988; *Williams et al.*, 2005).

**Table 2 Influence of heat treatment (85°C, 20 min) on mean particle diameter (z-average) and polydispersity index (PDI) of enzyme-treated (5 AU) native coacervates, denaturated WPI-pectin complexes and heated coacervates (0.5% w/w WPI, 0.25% w/w sugar beet pectin, 10 mM acetate buffer).**

| | native coacervates | | denaturated WPI-pectin complex | | heated coacervates | |
|---|---|---|---|---|---|---|
| pH | Z-Ave (nm) | PDI | Z-Ave (nm) | PDI | Z-Ave (nm) | PDI |
| 3 | 15490.5 ± 13413.8 | 0.825 ± 0.351 | 1882.0 ± 494.9 | 0.961 ± 0.078 | 10144.8 ± 1795.3 | 0.419 ± 0.223 |
| 4 | 199.9 ± 3.5 | 0.188 ± 0.004 | 475.1 ± 17.4 | 0.568 ± 0.120 | 245.8 ± 3.7 | 0.222 ± 0.017 |
| 5 | 247.1 ± 7.2 | 0.276 ± 0.010 | 451.3 ± 15.0 | 0.577 ± 0.033 | 271.2 ± 5.9 | 0.217 ± 0.015 |
| 6 | 307.7 ± 10.9 | 0.678 ± 0.087 | 183.2 ± 11.4 | 0.738 ± 0.147 | 272.1 ± 6.3 | 0.221 ± 0.013 |
| 7 | 284.5 ± 12.9 | 0.598 ± 0.122 | 181.0 ± 5.9 | 0.574 ± 0.126 | 276.8 ± 6.7 | 0.196 ± 0.019 |
| 8 | 316.7 ± 45.0 | 0.791 ± 0.082 | 181.0 ± 45.0 | 0.720 ± 0.123 | 273.0 ± 6.6 | 0.214 ± 0.011 |

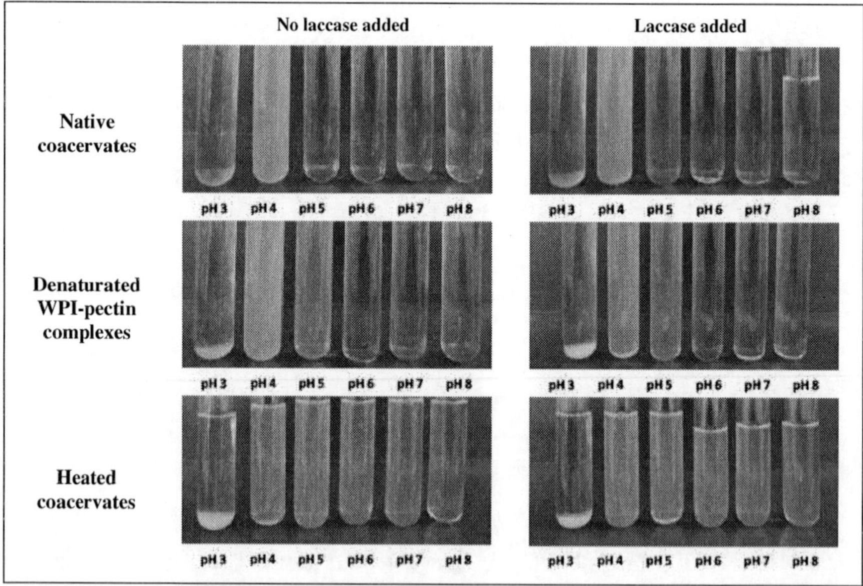


**Figure 3 Photographic images of native coacervates, laccase-treated native coacervates, denaturated WPI-beet pectin complexes, laccase-treated denaturated WPI-beet pectin complexes, heated coacervates, laccase-treated heated coacervates (0.5% w/w WPI, 0.25% w/w pectin) as a function of pH (3 to 8).**

The change in the particle diameter and polydispersity indices of the heated and unheated samples at pH 3 to 8 is shown in **Table 1**. The mean particle diameter of native coacervates increased slightly when the pH was increased from 4 to 8, indicating that the complexes dissociated. Moreover, the initial polydispersity index of the unheated complexes at pH 4 was 0.187 ± 0.014 nm, but increased steeply above 0.5, representing a very broad particle size distribution. Photographic images taken of all samples showed a decrease in turbidity with increasing pH from 4 to 8 (**Figure 3**), which further proves that dissociation of the protein-polysaccharide complexes occurred because both the WPI and the pectin molecules have high net negative charges leading to strong electrostatic repulsion forces (**Figure 2**) (*Jones et al.*, 2009; *Gentes et al.*, 2010). When biopolymer complexes containing only denaturated WPI and pectin were adjusted to different pH values, the coacervates also dissociated above the isoelectric point of the protein due to electrostatic repulsion (**Table 1**). On the other hand, the mean particle diameter and the polydispersity index of the heated complexes remained unchanged when the pH was adjusted to pH 8, indicating that the biopolymer particles remained large enough to scatter light (*Jones et al.*, 2009). Other studies reported similar effects when they heat-treated protein-polysaccharide complexes above the thermal denaturation temperature of the protein, thus suggesting that protein-protein bonds were favored to the detriment of protein-polysaccharide interactions (*Gentes et al.*, 2010; *Jones et al.*, 2010; *Jones et al.*, 2010). All complexes showed a pH-induced aggregation at pH 3 regardless of heat treatment, which could be attributed to unfavorable electrostatic interactions between WPI and pectin molecules (**Figure 2**).

**Influence of laccase on droplet characteristics of coacervates**

In this series of experiments, we determined the effect of pH on mean particle size, polydispersity indices and ζ-potential of differently heat-treated protein-polysaccharide-complexes to which laccase had been added. *Jones et al.* (2010) already proposed a core-shell structure of heat-treated biopolymer particles, where the core is comprised primarily of protein and the shell is formed by pectin molecules. Based on this structural organization, we hypothesized that laccase may promote covalent cross-links between beet pectin-beet pectin, WPI-WPI and/or WPI-beet pectin molecules via ferulic acid and/or tyrosine bonds. Cross-linked coacervates would, therefore, remain stable when the pH was shifted from 3 to 8, particularly at high pH values.

The effect of pH (3 - 8) on the mean particle diameter and polydispersity indices as well as ζ-potential of laccase-treated biopolymer particles are shown in **Table 2** and **Figure 4**, respectively. Laccase-treated coacervates showed similar particle sizes and polydispersity indices as non-enzymatically treated complexes at pH 4 (**Table 2**). Mean particle diameter and polydispersity indices of enzyme-treated complexes increased when the pH was increased from 4 to 8, whereas the turbidity decreased (**Figure 3**) indicating that laccase did not promote crosslinks between the biopolymers to further stabilize complex nanoparticles. In addition, an ITC experiment was carried out to further prove whether an enzyme-induced crosslinking between polysaccharide-polysaccharide, protein-protein, and/or protein-polysaccharide molecules took place. ITC technique is known to have a high sensitivity in order to quantify enzyme-substrate interactions (*Ladbury et al.*, 1996; *Blandamer et al.*, 1998). Ferulic acid, L-tyrosine, L-methionine, as well as their mixtures were chosen as substrates, which might be oxidized by the enzyme. Moreover, ferulic acid, L-tyrosine, and L-methionine are typical components to be found in sugar beet pectin or whey protein isolate, respectively (*Oosterveld et al.*, 1997; *Leroux et al.*, 2003; *Synytsya et al.*, 2003; *Chee et al.*, 2008; *Littoz et al.*, 2008).The enthalpy change resulting from injections of 1 mM solutions (10 mM acetate buffer, pH 4) of ferulic acid, L-tyrosine, L-methionine, and mixtures of ferulic acid and L-methionine, and ferulic acid and L-tyrosine, respectively, into laccase solutions was measured. Heat flow versus time profiles are shown **Figure 5**. Additionally, we calculated the average enthalpy change for a single injection; thereby, positive results indicate an exothermic change (**Table 3**).

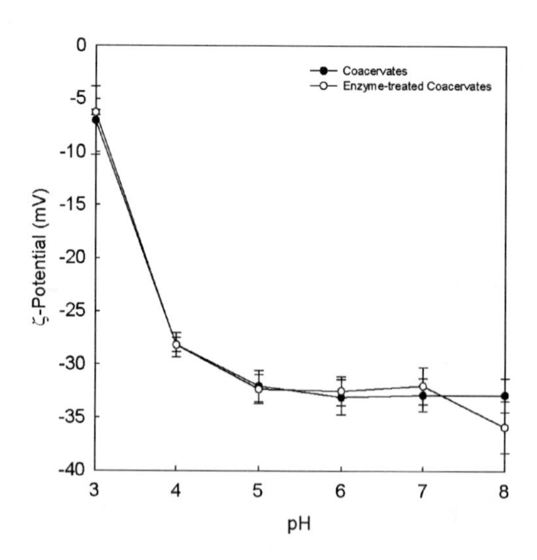


**Figure 4 Influence of pH (3 to 8) on the electrical properties (ζ-potential) of heated and enzyme-treated heated biopolymer complexes (0.5% w/w WPI, 0.25% w/w beet pectin, 10 mM acetate buffer).**

Our results indicated that the type of phenolic molecule affects the enzymatic oxidation by laccase (**Table 3**). In particular, an average enthalpy change of -943 ± 28 μJ (per injection) was measured when 1 mM ferulic acid was injected into the laccase solution (10 mM acetate buffer, pH 4). Moreover, no changes in enthalpy were observed after L-tyrosine or L-methionine were injected (**Figure 5**) indicating that laccase did not oxidize these amino acids. In addition, a decrease in enthalpy was observed when mixtures (ratio 1:1) of ferulic acid and L-tyrosine or ferulic and L-methionine were injected into the ITC measurement cell loaded with laccase which further proves that only an enzyme-induced oxidation of ferulic acid took place. Our results are in line with previously published studies (*Ma et al.*, 2011; *Chen et al.*, 2012). *Ma et al.* (2011) already demonstrated that unmodified WPI solutions without adding free phenolic components such as vanillic acid did not show any oxygen consumption even at higher enzyme dosages. Moreover, it was shown that laccase promotes only the crosslinking of beet pectin through oxidative coupling in WPI-beet pectin complexes (*Chen et al.*, 2012).

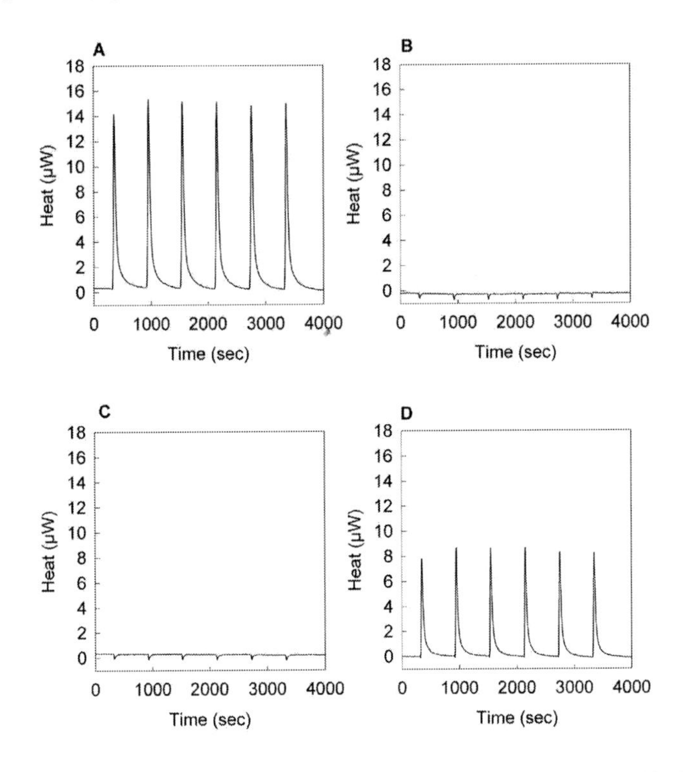


**Figure 5 Heat flow versus time profiles resulting from an injection of 10 μL aliquot containing 1 mM ferulic acid (A), 1 mM L-tyrosine (B), 1 mM L-methionine (C), and 1 mM ferulic acid-methionine (D) into a 996 μL titration cell filled with laccase solution (1 mg/10 ml).**

### D. Coacervate-coated primary emulsions

#### Influence of coacervate concentration on droplet characteristics

The change in the electrical charge of emulsions (0.5% (w/w) oil, 0.045% (w/w) WPI, pH 4) to which different concentrations of heated and non-heated coacervates (0 – 0.675% (w/w)) had been added was measured (**Figure 6**).

**Table 3 Enthalpy changes (per injection) of laccase-induced oxidation depending on substrate type (1 mM) (Blank -16 ± 5 μJ). Enzyme concentration 1 mg/10 ml.**

| substrate | enthalpy change (μJ) |
|---|---|
| ferulic acid | -943 ± 28 |
| L-tyrosine | 8 ± 11 |
| L-methionine | 16 ± 7 |
| ferulic acid/L-tyrosine | -428 ± 15 |
| ferulic acid/L-methionine | -491 ± 7 |

In the absence of coacervates, the electrical charge of primary emulsion was +46 ± 1 mV, indicating that the WPI in emulsion droplets was highly protonated at pH 4 (**Figure 6**). The electrical charge on the droplets became increasingly negative as coacervates were added to the emulsion suggesting that the negatively charged complexes adsorbed to the surface of the positively charged oil droplets forming a thick coacervate-WPI membrane (**Figure 6**). The ζ-potential became constant at a value of around -24 ± 1 mV when the coacervate concentration exceeded about 0.225% (w/w) indicating that the droplets became fully saturated with coacervates. For all subsequent experiments, a coacervate concentration of 0.375% (w/w) was used to prepare stable emulsions using a high shear blender to avoid droplet aggregation or flocculation. The mean particle diameter of primary emulsions increased after adsorption of coacervates from 229 ± 3 nm to 400 ± 28 nm. Static light scattering measurements showed monomodal distributed emulsions containing oil droplets stabilized by native or heat-treated coacervates.

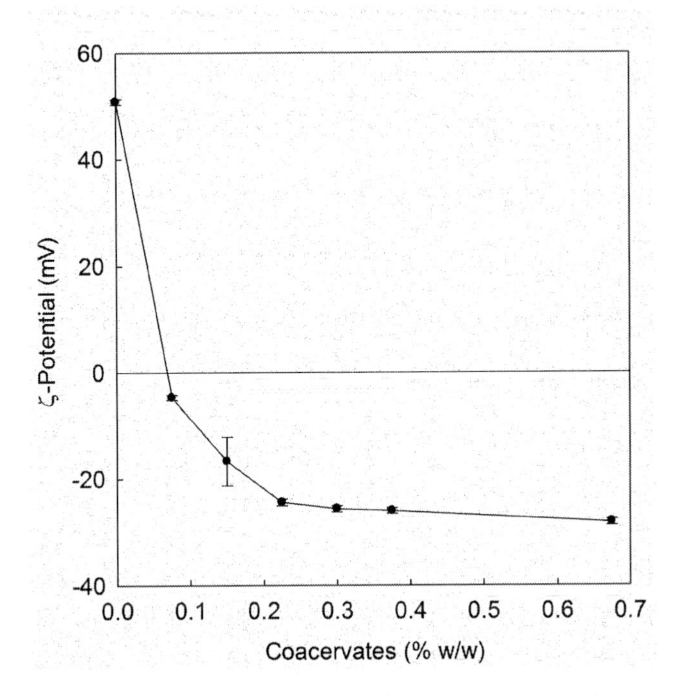


**Figure 6 Particle charge (ζ-potential) of WPI-stabilized emulsions as a function of added coacervates (0.5% w/w Miglyol, 0.045% w/w WPI, and 0-0.68% w/w coacervates, 10 mM acetate buffer, pH 4).**

## Effect of enzymatic cross-linking on pH- and freeze-thaw-stability

We hypothesized that coacervates that had been adsorbed to the surface of protein-coated oil droplets and then cross-linked would form tight membranes resisting harsh environmental conditions, such as changes in pH or temperature. In contrast, the coacervates which were not cross-linked were shown to detach, since charge repulsion between the pectin and WPI polymers occurs at high pH due to deprotonation of the WPI (**Figure 2**). In addition, the primary WPI-stabilized emulsions were prone to pH-induced flocculation, particularly when the pH was close to the p*I* of the protein, as already described in section A (**Figure 1**).

The effect of different pH on the particle size distributions of native and heated coacervate-coated primary emulsions was measured before and after laccase was added (**Figure 7**). Measurements of the particle size distributions of the coacervate-coated emulsions depending on pH showed that there was no change in size distributions with increasing pH. A heat treatment of the coacervates did not affect the pH stability of the coated primary emulsions. Thick and robust interfacial membranes have already been shown to improve the resistance of emulsions against environmental stresses (*Gu et al.*, 2004; *Aoki et al.*, 2005; *Gu et al.*, 2005; *Guzey et al.*, 2006).

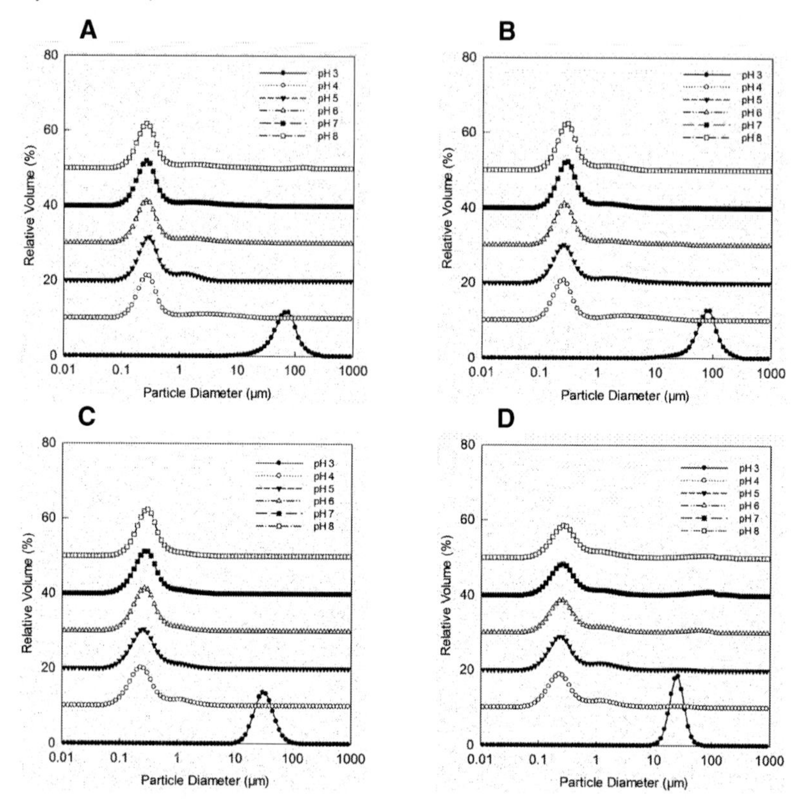


**Figure 7 Influence of pH on the particle size distribution of WPI-stabilized coated with native coacervates (A), laccase-treated native coacervates (B), heated coacervates (C), and laccase-treated heated coacervates (D) (0.5% w/w oil droplet concentration, 0.045% w/w WPI, 0.375% w/w coacervates).**

It is interesting to note that an addition of laccase to promote a cross-linking between adsorbed coacervates was not successful. Enzyme-treated emulsions showed similar particle size distributions when the pH was shifted from 4 to 8 (**Figure 7**). Moreover, enzyme-treated emulsions stabilized by nanosized biopolymer-particles showed extensive aggregation after two freeze-thaw cycles regardless of heat treatment (**Figure 8**). These results are in contrast to a study of Zeeb et al. published previously (*Zeeb et al.*, 2011). In this study, a laccase treatment increased the freeze-thaw stability of multilayered emulsions due to the formation of diferulic bonds between adsorbed pectin molecules, making the interfacial complex more resistant to rupture by oil and water crystals. Potentially, the structural arrangement of the pectin and protein molecules at the interface is thought to be a steric hindrance lowering the effectiveness of the laccase to oxidize ferulic acid groups, which further react to cross-linking between the polysaccharides. Ferulic acid on the sugar beet pectin is known to have hydrophobic properties favoring the interaction between non-polar patches on the protein (*Jones et al.*, 2009). Thus, this might sterically hinder the enzyme to cross-link the pectin. A laccase-induced linking of adsorbed pectin molecules was successful in comparison to multilaminar shells (**Figure 1**); it seems that the accessibility may play a key role in cross-linking the distinct substrate in an assembled complex structure. It was already shown that the rate of laccase-induced oxidation was slower for systems in which the emulsions had been homogenized together with the sugar beet pectin, as opposed to when prepared separately (*Zaidel et al.*, 2013). The author suggested that differences in the structural make-up of the emulsion system might have provided different availabilities of the ferulic acid groups attached to the pectin backbone for cross-linking. In addition, previous studies have demonstrated similar results for active enzymes assembled in three-dimensional nanostructures (*Rauf et al.*, 2006; *Kim et al.*, 2010). In these studies, the enzyme activity of the multilayer film mainly came from the outmost enzyme layer, and the authors concluded that substrates have a much lower accessibility for inner layers since they have to diffuse through the entire structure. In addition, the precise structural organization of the protein and polysaccharide molecules within the biopolymer particles formed is currently unknown, and the proposed core-shell structure mentioned above has not been verified (*Jones et al.*, 2009). Moreover, enzyme-treated emulsions, as well as untreated ones, showed extensive aggregation at pH 3, which might be attributed to the fact that this pH is close to the p$K_a$ of the pectin causing a lower attractive force between the biopolymers associated in the membrane (*Guzey et al.*, 2006; *Iwanaga et al.*, 2008).

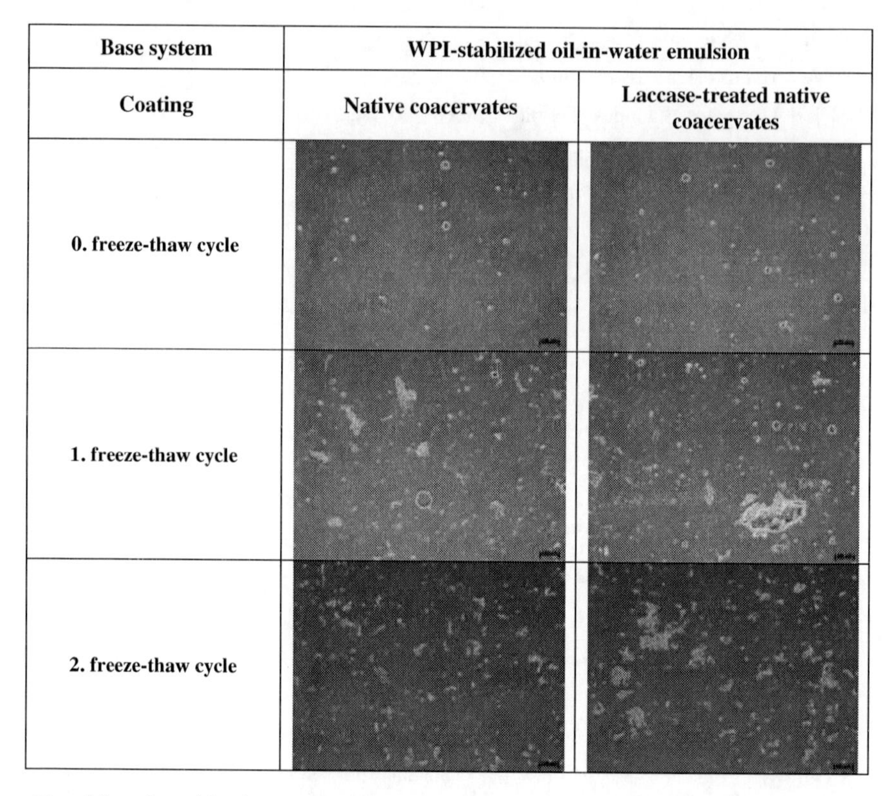


**Figure 8 Dependence of the microstructure on the number of freeze-thaw cycles (of WPI-stabilized coated with native coacervates and laccase-treated native coacervates emulsion droplets (0.5% w/w oil droplet concentration, 0.045% w/w WPI, 0.375% w/w coacervates). Emulsion samples were incubated in a -9°C water bath for 22 h and thawed by incubating in a 25°C water bath for 2 h (scale bar 100 μm).**

## Key insights

In summary, we could highlight some major effects obtained by our study:

- The formation of small-sized coacervates with particle sizes in the submicron range using complex coacervation method depends on the heat treatment of the polymer solutions.

- Laccase promotes an enzymatic crosslinking of polysaccharides, but does not induce oxidation of L-tyrosine or L-methionine, which could lead to the formation of covalent bonds between proteins and/or polysaccharides.

- Adsorption of native and heated coacervates using electrostatic depositioning technique improved the stability of protein-stabilized emulsions against pH-induced aggregation.

- A laccase-induced oxidation of adsorbed coacervates onto WPI-stabilized oil droplets was not observed, which might be due to spatial distribution of the molecules in the interface. In addition, only crosslinking of "non-buried" pectin layers seem to take place.

## CONCLUSION

This study has demonstrated that the assembly of biopolymers in preparing complex nano- and microstructures for foods plays a key role in terms of accessibility of the enzyme to its substrate. A targeted crosslinking of biopolymers in complex structures such as soluble complexes, coacervates, and hydrogel particles has to be induced before the assembly of the respective structures because the spatial distribution of molecules influences the enzymatic crosslinking. Moreover, the influence of membrane thickness on enzymatic crosslinking has a major impact on the arrangement of biopolymer based delivery systems. In contrast, an enzymatic crosslinking of multilaminar membranes in directed self-assembled structures such as multilayered emulsions is a useful tool to covalently crosslink adsorbed biopolymers. In the multilayered emulsions, an enzyme is not sterically hindered by already assembled structures.

## ACKNOWLEDGEMENTS

We would like to thank Herbstreith & Fox KG (Neuenbürg, Germany) as well as Fonterra GmbH (Hamburg, Germany) for generously providing us with polymer samples.

## REFERENCES


Aoki, T., E. A. Decker, et al. (2005): Influence of environmental stresses on stability of O/W emulsions containing droplets stabilized by multilayered membranes produced by a layer-by-layer electrostatic deposition technique. Food Hydrocolloids 19(2): 209-220.

Blandamer, M. J., P. M. Cullis, et al. (1998): Titration microcalorimetry. Journal of the Chemical Society - Faraday Transactions 94(16): 2261-2267.

Chanamai, R. and D. J. McClements (2002): Comparison of gum arabic, modified starch, and whey protein isolate as emulsifiers: Influence of pH, $CaCl_2$ and temperature. Journal of Food Science 67(1): 120-125.

Chanasattru, W., O. G. Jones, et al. (2009): Impact of cosolvents on formation and properties of biopolymer nanoparticles formed by heat treatment of β-lactoglobulin-pectin complexes. Food Hydrocolloids 23(8): 2450-2457.

Chee, K. S. and P. A. Williams (2008): Role of protein and ferulic acid in the emulsification properties of sugar beet pectin. Journal of Agricultural and Food Chemistry 56(11): 4164-4171.

Chen, B., H. Li, et al. (2012): Formation and microstructural characterization of whey protein isolate/beet pectin coacervations by laccase catalyzed cross-linking. LWT - Food Science and Technology 47(1): 31-38.

Dalgleish, D. G. and F. R. Hallett (1995): Dynamic light scattering: Applications to food systems. Food Research International 28(3): 181-193.

Demetriades, K., J. N. Coupland, et al. (1997): Physical properties of whey protein stabilized emulsions as related to pH and NaCl. Journal of Food Science 62(2): 342-347.

Dickinson, E. (1992): Faraday research article. Structure and composition of adsorbed protein layers and the relationship to emulsion stability. Journal of the Chemical Society, Faraday Transactions 88(20): 2973-2983.

Dickinson, E. (2010): Flocculation of protein-stabilized oil-in-water emulsions. Colloids and Surfaces B: Biointerfaces.

Gentes, M.-C., D. St-Gelais, et al. (2010): Stabilization of whey protein isolate-pectin complexes by heat. Journal of Agricultural and Food Chemistry 58(11): 7051-7058.

Gu, Y. S., A. E. Decker, et al. (2005): Production and characterization of oil-in-water emulsions containing droplets stabilized by multilayer membranes consisting of β-lactoglobulin, ι-carrageenan and gelatin. Langmuir 21(13): 5752-5760.

Gu, Y. S., E. A. Decker, et al. (2004): Influence of pH and ι-carrageenan concentration on physicochemical properties and stability of β-lactoglobulin-stabilized oil-in-water emulsions. Journal of Agricultural and Food Chemistry 52(11): 3626-3632.

Gu, Y. S., E. A. Decker, et al. (2007): Formation of colloidosomes by adsorption of small charged oil droplets onto the surface of large oppositely charged oil droplets. Food Hydrocolloids 21(4): 516-526.

Guzey, D. and D. McClements (2006): Influence of environmental stresses on o/w emulsions stabilized by β-lactoglobulin–pectin and β-lactoglobulin–pectin–chitosan membranes produced by the electrostatic layer-by-layer deposition technique. Food Biophysics 1(1): 30-40.

Guzey, D. and D. J. McClements (2006): Formation, stability and properties of multilayer emulsions for application in the food industry. Advances in Colloid and Interface Science 128-130: 227-248.

Hoffmann, M. A. M., S. P. F. M. Roefs, et al. (1996): Aggregation of β-lactoglobulin studied by in situ light scattering. Journal of Dairy Research 63(3): 423-440.

Hoffmann, M. A. M. and P. J. J. M. van Mil (1997): Heat-induced aggregation of β-lactoglobulin: Role of the free thiol group and disulfide bonds. Journal of Agricultural and Food Chemistry 45(8): 2942-2948.

Iwanaga, D., D. Gray, et al. (2008): Stabilization of soybean oil bodies using protective pectin coatings formed by electrostatic deposition. Journal of Agricultural and Food Chemistry 56(6): 2240-2245.

Jones, O. G., E. A. Decker, et al. (2009): Formation of biopolymer particles by thermal treatment of beta-lactoglobulin-pectin complexes. Food Hydrocolloids 23(5): 1312-1321.

Jones, O. G., U. Lesmes, et al. (2010): Effect of polysaccharide charge on formation and properties of biopolymer nanoparticles created by heat treatment of beta-lactoglobulin-pectin complexes. Food Hydrocolloids 24(4): 374-383.

Jones, O. G. and D. J. McClements (2008): Stability of biopolymer particles formed by heat treatment of β-lactoglobulin/beet pectin electrostatic complexes. Food Biophysics 3(2): 191-197.

Jones, O. G. and D. J. McClements (2010): Biopolymer nanoparticles from heat-treated electrostatic protein-polysaccharide complexes: Factors affecting particle characteristics. Journal of Food Science 75(2): N36-N43.

Kim, D. C., J. I. Sohn, et al. (2010): Controlled assembly for well-defined 3D bioarchitecture using two active enzymes. ACS Nano 4(3): 1580-1586.

Ladbury, J. E. and B. Z. Chowdhry (1996): Sensing the heat: the application of isothermal titration calorimetry to thermodynamic studies of biomolecular interactions. Chemistry & Biology 3(10): 791-801.

Leroux, J., V. Langendorff, et al. (2003): Emulsion stabilizing properties of pectin. Food Hydrocolloids 17(4): 455-462.

Littoz, F. and D. J. McClements (2008): Bio-mimetic approach to improving emulsion stability: Cross-linking adsorbed beet pectin layers using laccase. Food Hydrocolloids 22(7): 1203-1211.

Ma, H., P. Forssell, et al. (2011): Improving laccase catalyzed cross-linking of whey protein isolate and their application as emulsifiers. Journal of Agricultural and Food Chemistry 59(4): 1406-1414.

McClements, D. J. (2004): Food emulsions: Principles, practice, and techniques. Boca Raton, CRC Press.

Oosterveld, A., J. H. Grabber, et al. (1997): Formation of ferulic acid dehydrodimers through oxidative cross-linking of sugar beet pectin. Carbohydrate Research 300(2): 179-181.

Rauf, S., D. Zhou, et al. (2006): Building three-dimensional nanostructures with active enzymes by surface templated layer-by-layer assembly. Chemical Communications(16): 1721-1723.

Salminen, H. and J. Weiss (2013): Electrostatic adsorption and stability of whey protein–pectin complexes on emulsion interfaces. Food Hydrocolloids (0) 1-10; Article in Press.

Synytsya, A., J. Copiková, et al. (2003): Spectroscopic estimation of feruloyl groups in sugar beet pulp and pectin. International Sugar Journal 105(1258): 481-488.

Thibault, J. F. (1988): Characterisation and oxidative crosslinking of sugar-beet pectins extracted from cossettes and pulps under different conditions. Carbohydrate Polymers 8(3): 209-223.

Turgeon, S. L., C. Schmitt, et al. (2007): Protein-polysaccharide complexes and coacervates. Current Opinion in Colloid and Interface Science 12(4-5): 166-178.

Williams, P. A., C. Sayers, et al. (2005): Elucidation of the emulsification properties of sugar beet pectin. Journal of Agricultural and Food Chemistry 53(9): 3592-3597.

Zaidel, D. N. A., I. S. Chronakis, et al. (2013): Stabilization of oil-in-water emulsions by enzyme catalyzed oxidative gelation of sugar beet pectin. Food Hydrocolloids 30(1): 19-25.

Zeeb, B., L. Fischer, et al. (2011): Cross-linking of interfacial layers affects the salt and temperature stability of multilayered emulsions consisting of fish gelatin and sugar beet pectin. Journal of Agricultural and Food Chemistry 59(19): 10546-10555.

Zeeb, B., M. Gibis, et al. (2012): Crosslinking of interfacial layers in multilayered oil-in-water emulsions using laccase: Characterization and pH-stability. Food Hydrocolloids 27(1): 126-136.

# CHAPTER 7

## Influence of oil droplet concentration on transglutaminase-induced gelation of oil-in-water emulsions

*Benjamin Zeeb*[1], *Johanna Beicht*[1], *Thomas Eisele*[2], *Monika Gibis*[1],

*Lutz Fischer*[2], *Jochen Weiss*[1]

[1] Department of Food Physics and Meat Science, University of Hohenheim, Garbenstrasse 21/25, 70599 Stuttgart, Germany

[2] Department of Food Biotechnology, University of Hohenheim, Garbenstrasse 25,70599 Stuttgart, Germany

Reprinted from "*Transglutaminase-induced crosslinking of sodium caseinate stabilized oil droplets in oil-in-water emulsions*", Zeeb, B., Beicht, J., Eisele, T., Gibis, M., Fischer, L., Weiss, J., Food Research International, 2013, 53(2), p. 1712-1721 with permission from Elsevier.

**ABSTRACT**

The influence of oil volume fraction and sodium caseinate (Na-Cas) concentration of Na-Cas-stabilized oil-in-water emulsions on the ability of microbial transglutaminase to crosslink oil droplets was investigated. First, fine dispersed emulsions (5 – 60% (w/w) Miglyol; 2, 5, and 8% (w/w) aqueous Na-Cas) were prepared using a high pressure homogenizer. Second, microbial transglutaminase was added to Na-Cas-stabilized emulsions (37 °C, 15 h, pH 6.8) to initiate Na-Cas crosslinking. Texture profile analysis and rheological measurements indicated that at low protein concentrations droplet-droplet inter-crosslinking occurred above a critical oil volume concentration ($c_{oil} > 0.6$) yielding particle gels. There, strength of the droplet networks formed depended on oil droplet. At high Na-Cas contents, crosslinking of excess Na-Cas in the aqueous phase occurred and particle-filled continuous gels were formed instead. Theoretical calculations of mean distances between droplets and diffusion coefficients of droplets indicate that a certain probability of contact was required for microbial transglutaminase to be able to crosslink droplets.

**Keywords:** Oil-in-water emulsions; Gelation; Sodium caseinate; Transglutaminase; Protein network; Crosslinking

## INTRODUCTION

Many foods, pharmaceuticals and personal care products are natural or manufactured emulsion-based materials with complex structures that provide specific functional properties required for their use *e.g.* rheology, optical appearance, mechanical and chemical stability, or bioactivity (*Macierzanka et al.*, 2011). Emulsions consist of two or more partially or completely immiscible liquids with one liquid being dispersed in the other in the form of small droplets (*Walstra et al.*, 2003; *McClements*, 2004). Droplets in emulsions are typically stabilized by a layer of small molecule surfactants or surface active polymers (*e.g.* proteins, protein-carbohydrate conjugates, or block copolymers) adsorbed at the oil-water interface (*McClements*, 2004; *McClements*, 2004; *Dickinson*, 2012). Moreover, droplet characteristics *e.g.* droplet concentration, droplet size, droplet charge and interfacial composition govern the macroscopic behaviour of emulsions (*Dickinson*, 2012). At high droplet concentrations, emulsions may attain a viscoelastic behaviour due to the droplets being closely packed. There, droplets are not chemically linked, but are in immediate contact with each other able to transfer stresses across the droplet network. As a consequence, highly concentrated emulsions have an elastic modulus and a yield stress, both properties that are often required to provide texture to foods, adhesion to paints, or efficacy to topical medications. For such networks to form, oil volume fractions must typically exceed 75% and the emulsions is consequently mainly composed of oil (*Mezzenga et al.*, 2010).

Alternatively, the stabilizing polymers in droplet interfaces may be physically or chemically crosslinked to form a network of droplets. Such systems are referred to as emulsion gels. Of the above mentioned surface-active polymers, proteins are particularly well suited to generate emulsion gels since they possess a wide variety of functional groups that can be covalently crosslinked (*Lee et al.*, 2006; *Tang et al.*, 2011). In addition to providing the above mentioned functional properties, emulsion gels have shown to be suitable delivery systems for a variety of biologically active compounds such as carotenoids, polyunsaturated fatty acids, phytosterols and vitamins, and have thus of late attracted considerable interest (*Chen et al.*, 2006; *Velikov et al.*, 2008; *McClements et al.*, 2009).

The formation of protein-emulsion droplet networks may be achieved by means of heat treatment, acidification with glucono-δ-lactone (GDL), addition of coagulants such as divalent salts (e.g. $CaCl_2$) or biopolymers (*Tang et al.*, 2011; *Yang et al.*, 2011; *Dickinson*, 2012).

Moreover, oxidative enzymes including microbial peroxidase, fungal laccase, and bovine plasma monoamine oxidase as well as transglutaminase may be used to chemically crosslink droplets (*Dickinson*, 1997; *Tang et al.*, 2011; *Yang et al.*, 2011; *Dickinson*, 2012). Gels formed by non-thermal treatments are referred to as cold set gels and have shown to be particularly well suited to deliver heat-labile bioactives and nutraceuticals (*Tang et al.*, 2011; *Yang et al.*, 2011). Generally, one can distinguish between two structural arrangements in such systems depending on protein and oil droplet concentration: (i) emulsions-filled protein gels in which oil droplets are embedded in a crosslinked protein matrix, and (ii) protein-stabilized emulsion droplet gels in which droplets have been crosslinked (*Dickinson*, 2012). To date, it is not yet completely clear under which conditions one or the other is formed, particularly if a cold setting method such as enzymatic crosslinking is used.

Application of enzymes is considered to be a mild processing technology and can be implemented using conventional processing technologies such as mixing or homogenization (*Dickinson*, 1997; *Minussi et al.*, 2002; *Zeeb et al.*, 2011). Thus, several biomimetic approaches have been carried out to study the ability of enzymes to crosslink biopolymers such as proteins or polysaccharides (*Dickinson*, 1997; *Littoz et al.*, 2008; *Zeeb et al.*, 2011; *Zeeb et al.*, 2012). The most popular enzyme to promote cold-set protein gelation in foods is microbial transglutaminase. This enzyme catalyses the acyl transfer between glutamine and lysine residues in proteins resulting in the formation of both inter- and intramolecular covalent bonds (*Motoki et al.*, 1998; *De Jong et al.*, 2002). Recently, we conducted a series of studies to crosslink biopolymer layers in multilayered interfaces using enzymes. We demonstrated that application of the oxidoreductase laccase led to formation of crosslinks between beet pectins in a fish gelatin-beet pectin multilayered oil-in-water emulsion, thereby improving its salt, pH, heat, and freeze-thaw stability (*Littoz et al.*, 2008; *Zeeb et al.*, 2011; *Zeeb et al.*, 2012). Moreover, we showed that a diffusion-driven destabilization of *n*-alkane-in-water emulsions via Ostwald ripening could be retarded by enzymatically crosslinking the interfacial membranes surrounding the oil droplets (*Zeeb et al.*, 2012).

The objective of the present study was to establish a better understanding of the mechanism of action of a crosslinking enzyme added to protein-stabilized oil-in-water emulsions varying in droplet concentration and protein content. We were particularly interested in finding out under which conditions *intra-interfacial* crosslinking *vs. interdroplet* crosslinking occurred. To this purpose, a series of oil-in-water emulsions with various oil volume fractions stabilized by sodium caseinate were prepared, and treated with transglutaminase. Sodium caseinate is a mixture of $\alpha_{s1}$-, $\alpha_{s2}$-, $\beta$-, and $\kappa$-casein with approximately 75% of the casein being composed of $\alpha_{s1}$- and $\beta$-casein (*Sanchez et al.*, 2005; *Lee et al.*, 2006). Caseinates are known to have a disordered and flexible structure and therefore tend to be more readily crosslinked by transglutaminase than globular and more compact proteins such as *e.g.* whey proteins (*Dickinson et al.*, 1996; *Motoki et al.*, 1998; *Sharma et al.*, 2002; *Hinz et al.*, 2007). We hypothesized that a critical oil droplet concentration may be required to promote droplet-droplet crosslinking since the probability of contact governed by the mean surface distance between particles increases as the oil droplet concentration increases. All emulsions were prepared at neutral pH.

## MATERIALS AND METHODS

**Materials.** Sodium caseinate (Na-Cas, #L080512201) was purchased from Rovita GmbH (Engelsberg, Germany). As per manufacturers specification, caseinate contained $\geq$ 88% protein, $\leq$ 6% moisture, $\leq$ 4.5% ash, $\leq$ 1.5% fat, and $\leq$ 1% lactose. Sodium caseinate was used without further purification. Miglyol 812N, a medium chain triacylglyceride mixture, was obtained from Sassol Germany GmbH (Brunsbüttel, Germany). It served as a model lipid in the oil-in-water emulsion. Transglutaminase (TGase) was obtained from Ajinomoto Foods Europe SAS (Hamburg, Germany). Z-Gln-Gly (Z-L-Glutaminyl-Glycine, #C6154), Iron (III) chloride ($FeCl_3$, #7705-08-0), and trichloroacetic acid (TCA, #76-03-9) were obtained from Sigma-Aldrich Co. (Schnelldorf, Germany). Calcium chloride ($CaCl_2$, #CN93.1, purity $\geq$ 99.0%) was purchased from Carl Roth GmbH & Co. KG (Karlsruhe, Germany). Analytical grade hydrochloric acid (HCl) and sodium hydroxide (NaOH) were purchased from Carl Roth GmbH & Co. KG (Karlsruhe, Germany). Double-distilled water was used in the preparation of all samples.

**Transglutaminase activity.** Transglutaminase activity was determined according to the method of *Folk et al.* (1966) with some modifications using Z-Gln-Gly as a substrate. A mixture of 12 mg/ml Z-Gln-Gly, 100 mM hydroxylamine, 10 mM glutathione (reduced) and 5 mM $CaCl_2$ was prepared in Tris buffer (200 mM, pH 6.0). The reaction cocktail was incubated at 37 °C for 5 min in a thermo mixer. 30 µl enzyme solution (100 mg/ml) was added to initiate the reaction. The reaction was stopped by addition of TCA (12% (w/v), 500 µl) after 10 min. Finally, a $FeCl_3$ solution (5% (w/v), 500 µl) prepared in hydrochloric acid (100 mM) was added to the solution and the absorbance was spectrophotometrically measured at 525 nm (GE, München, Germany). One katal of transglutaminase activity was defined as the amount of enzyme required to form 1 mol γ-monohydroxamate per second at 37 °C and pH 6.0. A transglutaminase activity of 73.1 nkat/ml was measured.

**Solution preparation.** An aqueous emulsifier solution was prepared by dispersing 2, 5, and 8% (w/w) sodium caseinate powder in double distilled water containing sodium azide (0.02% (w/w)) as an antimicrobial agent. All solutions were stirred at ambient temperature over night to ensure complete hydration and then adjusted to a pH of 6.8 using 1 M HCl and/or 1 M NaOH.

**Emulsion preparation.** Protein concentrations in the continuous phase were kept constant at 2, 5, and 8% (w/w), respectively, whereas the oil droplet concentration was varied between 5 and 60% (w/w). Na-Cas dispersions were mixed with Miglyol using a high shear blender (Standard Unit, IKA Werk GmbH, Germany) for 2 min. The coarse premixes were then passed through a high pressure homogenizer (Avestin, Inc., Ottawa, Ontario, Canada) three times at 1000 bar. Oil-in-water emulsions with higher oil droplet concentrations (50 – 70% w/w) were prepared by blending oil and emulsifier solutions with a high shear homogenizer (Silent Crusher, Heidolph Instruments GmbH and Co. KG, Schwabach, Germany) for 3 min at 20000 rpm.

**Enzymatic treatment.** Caseinate dispersions or emulsions were treated with transglutaminase by mixing them with powdered enzyme (73.1 μkat/l) using a vortexer. A ratio of 1 mg enzyme per 1 g protein solution or emulsion was sufficient to promote crosslinks. Samples were transferred to sealed glass cylinders (height: 100 mm, diameter: 30 mm) and incubated at 37 °C using a temperature controlled cabinet for 15 h and then cooled in a water bath to room temperature. All test tubes inversely shown indicate a sol-gel transition after incubation with transglutaminase.

**Droplet size distributions.** Droplet size distributions were determined by static light scattering (Horiba LA-950, Retsch Technology GmbH, Haan, Germany). Samples were withdrawn and diluted to a droplet concentration of approximately 0.005% (w/w) with water (pH 6.8) to prevent multiple scattering effects. The instrument measures the angular dependence of the intensity of the laser beam scattered by the dilute emulsions and then uses the Mie theory to calculate the droplet size distributions that gave the best fit between theoretical predictions and empirical measurements. A refractive index ratio of 1.08 (ratio of the indices between the oil and water phase) was used.

**Rheological properties.** The rheological properties of samples with and without transglutaminase were determined using a modular compact oscillatory rheometer (MCR 300, Anton Paar, Stuttgart, Germany). The rheometer was equipped with a coaxial cylinder (CC-27, cup diameter: 28.92 mm, bob diameter: 26.66 mm). Samples were subjected to a frequency sweep from 0.1 to 100 Hz at a strain of 0.1%. Storage modulus ($G'$), loss modulus ($G''$), complex viscosity ($\eta^*$), and loss tangent (tan $\delta$) were recorded. All measurements were carried out at 25 °C.

**Texture profile analysis.** Samples (20 g, height: 40 mm, diameter: 25 mm) that had gone through a sol-gel transition were subjected to a probe penetration (fracture) test using a TA-XT2 texture analyser (Stable Micro System, London, UK) equipped with a circular penetration probe (diameter: 10 mm). The probe was advanced at a probe speed of 1 mm/s. The probe penetrated the gels to a maximum depth of 15 mm. All measurements were carried out at room temperatures using 8 samples, and mean values and standard deviations for the maximum force (N) were calculated using the TA-XT2 texture exponent software.

**Statistical analysis.** All experiments were repeated at least 2 times using freshly prepared samples. Means and standard deviations were calculated from a minimum of three measurements using Excel.

## RESULTS AND DISCUSSION

### Characterization and enzymatic treatment of Na-Cas dispersions

We examined the rheological behavior of protein dispersions before and after addition of transglutaminase. Therefore, sodium caseinate solutions with various protein concentrations ranging from 1 to 10% (w/w) were dissolved in distilled water (pH 6.8), treated with transglutaminase and incubated at 37 °C for 15 h. **Figure 1** shows the appearance and the complex viscosity of sodium caseinate solutions before and after addition of transglutaminase indicating that crosslinking occurred above a critical Na-Cas concentration (> 7% (w/w)) under the prevalent conditions. This is in agreement with earlier studies that have also shown that transglutaminase induced formation of crosslinks of proteins within a reasonable timeframe at elevated temperatures (*Nonaka et al.*, 1992; *Dickinson et al.*, 1996; *Færgemand et al.*, 1997; *Færgemand et al.*, 1999). *Kellerby et al.* (2006) found that during a 16 h treatment of caseinate-stabilized oil-in-water emulsions at 40 °C, ammonia was released for approximately 12 h after addition of the enzyme. Higher protein concentrations led to the formation of a firm protein network whose viscosity could not be measured without breaking down the structure. Below a critical protein concentration (< 7% (w/w)) samples remained fluid - a fact that might be explained by the mechanism of action of the enzyme itself (**Figure 1B**). Transglutaminase catalyses the formation of glutamine and lysine residues in proteins; however, in the absence of sufficient high concentrations of lysine residues or other primary amines, water reacts as a nucleophile which results in a deamidation of glutamines and thus no crosslinks are formed (*Kuraishi et al.*, 2001; *De Jong et al.*, 2002; *Sharma et al.*, 2002; *Partanen et al.*, 2009).

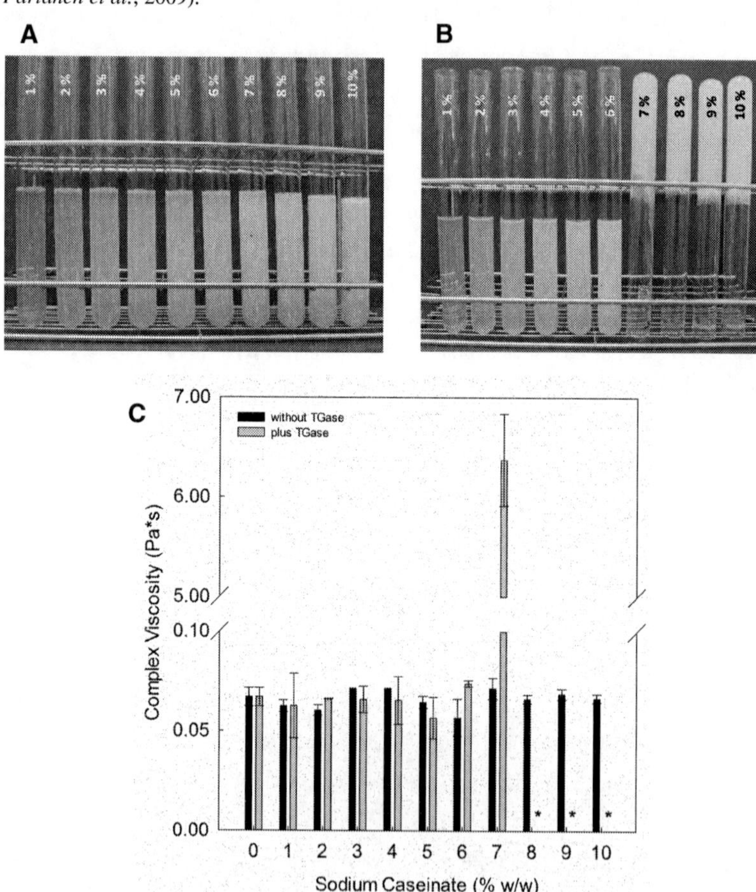


**Figure 1 Photographic pictures of sodium caseinate solutions before (A) and after (B) transglutaminase-induced crosslinking as well as their complex viscosities (C) (pH 6.8) (* indicates complex viscosities approaching infinity).**

**Characterization and enzymatic treatment of Na-Cas-stabilized emulsions**

High pressure homogenization was used to prepare oil-in-water emulsions with various oil contents (5 – 60% (w/w)). We also varied the emulsifier concentration (2, 5, and 8% (w/w)) to investigate its influence before and after addition of transglutaminase. The volume fraction $\Phi$ the oil content, and the mean particle diameter $d_{32}$ of all emulsion samples prepared are shown in **Table 1**. The volume fraction $\Phi$ can be calculated as:

$$\phi = \frac{V_{oil}}{V_{emulsion}} = \frac{V_{oil}}{V_{oil}+V_{water}} \tag{1}$$

where $V_{oil} = m_{oil}/\rho_{oil}$ and $V_{water} = m_{water}/\rho_{water}$ using different densities for the water phase ($\rho_{water}$ = 1000 kg/m$^3$) and the oil phase ($\rho_{oil}$ = 950 kg/m$^3$). The mean particle diameter $d_{32}$ was calculated from the droplet size distribution as:

$$d_{32} = \frac{\sum d_i^3 n_i}{\sum d_i^2 n_i} \tag{2}$$

where $n_i$ is the number of particles in a class having a diameter of $d_i$.

**Table 1 Oil volume fraction ($\Phi$) calculated by equation (1) and mean surface diameter ($d_{32}$) of oil-in-water emulsions stabilized by various sodium caseinate concentrations (w/w).**

| oil (%) | Φ (-) | 2% Na-Cas $d_{32}$ (µm) | 5% Na-Cas $d_{32}$ (µm) | 8% Na-Cas $d_{32}$ (µm) |
|---|---|---|---|---|
| 5 | 0.0525 | 0.188 ± 0.001 | 0.187 ± 0.000 | 0.289 ± 0.046 |
| 10 | 0.1047 | 0.202 ± 0.005 | 0.171 ± 0.002 | 0.227 ± 0.039 |
| 15 | 0.1567 | 0.198 ± 0.008 | 0.217 ± 0.005 | 0.164 ± 0.000 |
| 20 | 0.2083 | 0.192 ± 0.004 | 0.182 ± 0.002 | 0.237 ± 0.000 |
| 25 | 0.2597 | 0.204 ± 0.001 | 0.203 ± 0.000 | 0.356 ± 0.086 |
| 30 | 0.3109 | 0.236 ± 0.002 | 0.237 ± 0.000 | 0.329 ± 0.107 |
| 35 | 0.3618 | 0.310 ± 0.002 | 0.233 ± 0.000 | 0.236 ± 0.001 |
| 40 | 0.4124 | 0.312 ± 0.001 | 0.240 ± 0.001 | 0.250 ± 0.001 |
| 45 | 0.4627 | 0.365 ± 0.001 | 0.242 ± 0.001 | 0.258 ± 0.000 |
| 50 | 0.5128 | 0.472 ± 0.002 | 0.231 ± 0.000 | 0.291 ± 0.014 |
| 55 | 0.5627 | 0.667 ± 0.011 | 0.313 ± 0.001 | 0.289 ± 0.001 |
| 60 | 0.6122 | 0.911 ± 0.007 | 0.339 ± 0.003 | 0.316 ± 0.001 |

A statistical analysis of the results showed that the mean particle diameter $d_{32}$ remained constant at Na-Cas concentration of at oil contents below 30% (w/w) (**Table 1**). Above, mean particle diameters increased with increasing oil droplet concentrations in all emulsions, but particularly in those that had been made with a low Na-Cas content (2% (w/w)). The results are in line with expectations (*Dickinson*, 2001; *McClements*, 2004; *Sanchez et al.*, 2005). At low oil droplet contents, the amount of available Na-Cas was sufficient to stabilize the droplets that are generated via the shear forces exhibited under the specific homogenization conditions. Hence, all emulsions had about the same mean droplet size. At higher oil contents, the interfacial area becomes too high and can no longer be completely covered, particularly at low emulsifier concentrations leading to an increase in mean droplet size (*Dickinson*, 2001; *Sanchez et al.*, 2005). This behaviour is ultimately governed by the amount of free versus adsorbed Na-Cas and a more detailed consideration of these concentrations can be found later in the theoretical section.

**Table 2 Changes in storage (*G´*) and loss modulus (*G´´*) before and after addition of transglutaminase (TGase) to oil-in-water emulsions stabilized by 5% (w/w) sodium caseinate at an oscillation frequency of 1.1 Hz.**

| | without TGase | | with TGase | |
|---|---|---|---|---|
| **oil (%)** | **storage modulus (Pa)** | **loss modulus (Pa)** | **storage modulus (Pa)** | **loss modulus (Pa)** |
| 5 | 0 | 0.486 ± 0.026 | 0 | 0.496 ± 0.030 |
| 10 | 0 | 0.465 ± 0.042 | 0 | 0.488 ± 0.004 |
| 15 | 0 | 0.411 ± 0.093 | 0 | 0.438 ± 0.035 |
| 20 | 0 | 0.505 ± 0.049 | 0 | 0.447 ± 0.048 |
| 25 | 0 | 0.478 ± 0.011 | 0 | 0.468 ± 0.033 |
| 30 | 0 | 0.476 ± 0.026 | 0 | 0.418 ± 0.012 |
| 35 | 0.283 ± 0.400 | 0.439 ± 0.333 | 0.372 ± 0.525 | 0.449 ± 0.078 |
| 40 | 2.600 ± 0.156 | 5.005 ± 0.035 | 0.575 ± 0.537 | 1.775 ± 0.417 |
| 45 | 38.4 ± 8.2 | 11.9 ± 1.6 | 20.7 ± 3.0 | 8.2 ± 0.8 |
| 50 | 72.6 ± 27.7 | 14.0 ± 1.1 | 40.5 ± 5.0 | 7.6 ± 1.5 |
| 55 | 410.0 ± 287.1 | 43.1 ± 29.3 | 539.5 ± 454.7 | 32.4 ± 23.0 |
| 60 | 1745.0 ± 912.2 | 206.0 ± 116.0 | 4240.0 ± 1555.6 | 257.0 ± 86.3 |

Results of the dynamic rheological measurements show that the mechanical behaviour of the emulsions is predominately viscous below a critical concentration and becomes increasingly elastic at high oil droplet contents (**Figure 2**) (*Chanamai et al.*, 2000; *McClements*, 2004; *Dickinson*, 2012). At oil contents of 5 – 30% (w/w) the particles are sufficiently far apart and do not interact with each other. Their movement is mainly determined by Brownian motion (*McClements*, 2004). Increases in the viscosity of the emulsions are solely due to droplet-induced deviations in the laminar flow pattern (*Dickinson*, 2012). Moreover, increases in loss and storage moduli at higher oil droplet concentrations indicate that particles begin to interact appreciably with each other. Their free movement is hindered (*McClements*, 2004). At very high droplet concentrations, the droplets are very closely packed able to transmit stresses across the droplet network (*Chanamai et al.*, 2000; *McClements*, 2004). **Figure 2** also indicates that the rheological properties of all emulsions depended on the total caseinate content. Both loss modulus ($G''$) and the storage modulus ($G'$) increased with increasing protein concentrations, especially at oil droplet concentrations > 30% (w/w) indicating that emulsions became more solid-like (*Chen et al.*, 1998; *Dickinson*, 2012).

Rheological measurements and texture profile analysis was then used to measure changes in texture and viscoelastic parameters ($G'$ and $G''$) of emulsions that were treated with transglutaminase at 37 °C for 15 h. As an example, **Table 2** shows results of measurements of $G'$ and $G''$ for emulsions stabilized by 5% (w/w) Na-Cas and having oil contents from 5 to 60% (w/w) before and after addition of transglutaminase. **Figure 3** shows photographic images of test tubes before (**Figure 3A**) and after enzymatic treatment (**Figure 3B**) as well as results of force penetration measurements (**Figure 3C**) for emulsions containing 8% (w/w) Na-Cas and 5 - 60% (w/w) oil. The data indicate that the enzyme did alter the properties and behaviour of emulsions containing 2 or 5% (w/w) Na-Cas at oil droplet contents above 50% (w/w) (*Lee et al.*, 2006; *Tang et al.*, 2011; *Yang et al.*, 2011). Below 50% (w/w) oil content, neither droplet-droplet crosslinking nor crosslinking of excess Na-Cas in the aqueous phase occurred since no significant differences in loss or storage modulus after addition and incubation of transglutaminase was measured. For droplet-droplet crosslinking droplets were likely not sufficiently closely packed while for the latter the protein concentration was too low (**Figure 1**) (*Kuraishi et al.*, 2001; *De Jong et al.*, 2002).

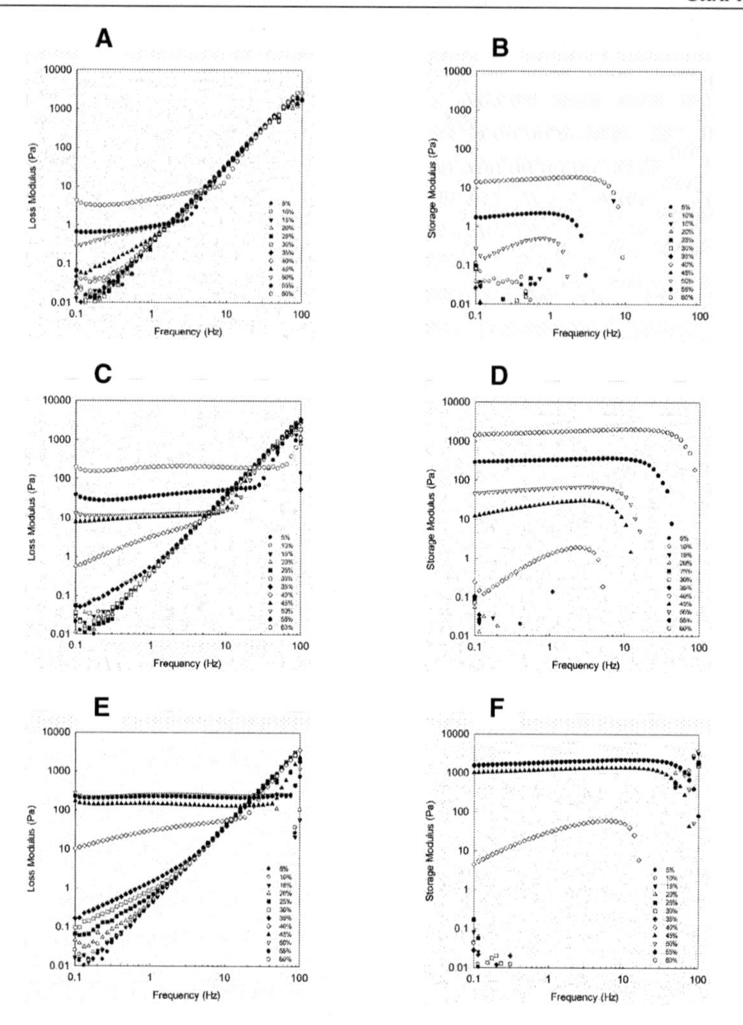


**Figure 2 Frequency dependence of storage modulus and loss modulus of protein-stabilized emulsions at various oil fractions (5 - 60% (w/w)) as a function of sodium caseinate concentration: 2 (A, B), 5 (C, D), 8 (E, F) % (w/w).**

In contrast, all emulsions containing 8% (w/w) sodium caseinate in the aqueous phase underwent a sol-gel transition when transglutaminase was added (*Dickinson*, 2012). Penetration measurements indicated a peak of maximum force observed at 20% (w/w) oil content, whereas gel strength decreased further with increasing oil droplet concentration. Studies conducted previously had shown that addition of transglutaminase generally led to formation of emulsion-filled protein-gels (*Sharma et al.*, 2002; *Lee et al.*, 2006; *Tang et al.*, 2011; *Yang et al.*, 2011) and that increasing oil volume fraction increased their strength. Such systems were found to better retain aroma compounds upon storage than emulsions not treated with enzymes (*Lee et al.*, 2006). The discrepancies to results in this study may be attributed to the fact that the studies were conducted with high soy protein isolate concentrations so that excess protein was likely present in the aqueous phase that could then be crosslinked to form a protein network in which the droplets were suspended.

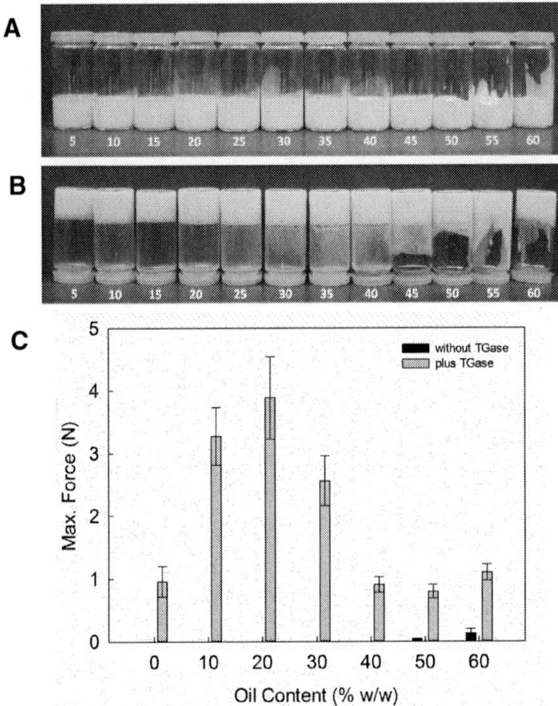


**Figure 3 Photographic images of oil-in-water emulsions (5 – 60% (w/w)) stabilized by 8% (w/w) sodium caseinate before (A) and after (B) transglutaminase addition as well as fracture force (C) of emulsions (10 - 60% (w/w), 8% (w/w) Na-Cas) after transglutaminase was added.**

In contrast, our results suggest if concentrations of excess protein were insufficient, a continuous Na-Cas network could not be formed, and that transglutaminase-induced droplet-droplet crosslinking only occurred when the oil volume fraction exceeds a critical value (**Figure 4**). To further assess this apparent transition from a droplet network to protein network containing embedded droplets, a high shear blender was used to prepare an additional series of emulsions that contained oil in excess of 60% (w/w), namely from 50 – 70% (w/w).

**Table 3 Texture profile analysis of 60% (w/w) oil-in-water emulsions stabilized by various sodium caseinate concentrations (w/w) before and after addition of transglutaminase (37 °C, 15 h, pH 6.8). All samples were prepared using a high shear blender (20000 rpm, 3 min).**

| oil (%) | 2% Na-Cas max. force (N) | | 5% Na-Cas max. force (N) | | 8% Na-Cas max. force (N) | |
|---|---|---|---|---|---|---|
| | without TGase | with TGase | without TGase | with TGase | without TGase | with TGase |
| 50 | 0.040 ± 0.003 | 0.042 ± 0.004 | 0.120 ± 0.055 | 0.126 ± 0.072 | 0.200 ± 0.045 | 4.727 ± 0.600 |
| 55 | 0.041 ± 0.002 | 0.039 ± 0.002 | 0.145 ± 0.018 | 0.152 ± 0.011 | 0.230 ± 0.035 | 5.775 ± 0.626 |
| 60 | 0.042 ± 0.003 | 0.041 ± 0.003 | 0.152 ± 0.029 | 0.205 ± 0.027 | 0.244 ± 0.031 | 5.704 ± 1.772 |
| 65 | 0.041 ± 0.017 | 0.040 ± 0.003 | 0.155 ± 0.080 | 0.419 ± 0.048 | 0.257 ± 0.167 | 4.536 ± 1.563 |
| 70 | 0.100 ± 0.029 | 0.106 ± 0.029 | 0.160 ± 0.020 | 0.593 ± 0.120 | 0.276 ± 0.1230 | 3.643 ± 0.847 |

Preliminary experiments demonstrated that, 60% (w/w) was the highest concentration we were able to manufacture emulsions by high pressure homogenization and thus we had to use another homogenization method. High shear blender are widely used to generate emulsions with high oil volume fractions, whereas high pressure valve homogenizers are more suitable for low and intermediate viscosity materials to achieve small droplet sizes (*McClements*, 2004). The mean particle diameter of these emulsions were higher than emulsion prepared by high pressure homogenization (**Figure 4A**) which may have influenced droplet packing. Nevertheless, an enzyme-induced droplet-droplet gel could now be formed at droplet concentrations exceeding 65 or 60% (w/w) in emulsions containing 2 or 5% (w/w) Na-Cas, respectively (**Figure 4E, 4F and 4G**). The results further confirm that apparently a certain number of particles is needed to induce particle gels (*Lee et al.*, 2006; *Tang et al.*, 2011). However, all emulsion samples stabilized by 2% (w/w) Na-Cas showed extensive coalescences and creaming after incubation with transglutaminase indicating that there was not sufficient emulsifier present to completely cover the oil-in-water interface (**Figure 4E**). Fractures studies on the texture analyser provide further proof for droplet-droplet crosslinking at 2 and 5% (w/w) Na-Cas versus excess protein crosslinking at 8% (w/w) Na-Cas after enzyme addition (**Table 3**).

**Theoretical considerations**

Results above shown that depending on oil droplet and protein content, addition of transglutaminase may (i) not lead to the formation of a continuous network (ii) lead to the formation of a droplet network or (iii) lead to the formation of a protein network with embedded droplets. To better understand these different outcomes, we carried out some calculations in which we attempted to relate droplet concentration to average distances and movements of droplets since a critical mean surface distance $h$ between dispersed oil droplets may be crucial to promote inter droplet crosslinking. The following assumptions were made to carry out the calculations: (i) the particles dispersed in the continuous phase are spherical and not deformable; (ii) all particles are evenly and monomodal distributed through the aqueous phase; (iii) in concentrated emulsions particles arrange in geometrical ideal structures referred as close packing (*Walstra et al.*, 2003; *McClements*, 2004); (iv) the enzyme transglutaminase is not affected by repulsive forces acting between oil droplets due to the presence of the charged Na-Cas. Then, the mean surface distance $h$ as a function of oil volume fraction ($\Phi$) and particle diameter ($d$) can be calculated by the Woodcock equation (*Woodcock*, 1987):

$$\frac{h}{d} = \sqrt{\frac{1}{3\pi\Phi} + \frac{5}{6}} - 1 \tag{3}$$

Moreover, if the position of an oil droplet is within the centre of a packing arrangement, then the relationship between mean surface distance, oil volume fraction, and particle diameter is given as (*Kamiya et al.*, 2010):

$$\frac{h}{d} = \sqrt[3]{\frac{\pi}{3\Phi\sqrt{2}}} - 1 \tag{4}$$

By using the above equations (3) and (4), we were able to calculate the mean surface distance depending on packing structure for a specific mean particle diameter (**Figure 5**). Three excesses of particle diameters were selected, namely 30, 300, and 3000 nm, respectively, to properly visualize the differences in mean surface distances as a function of oil volume fraction and droplet packing.

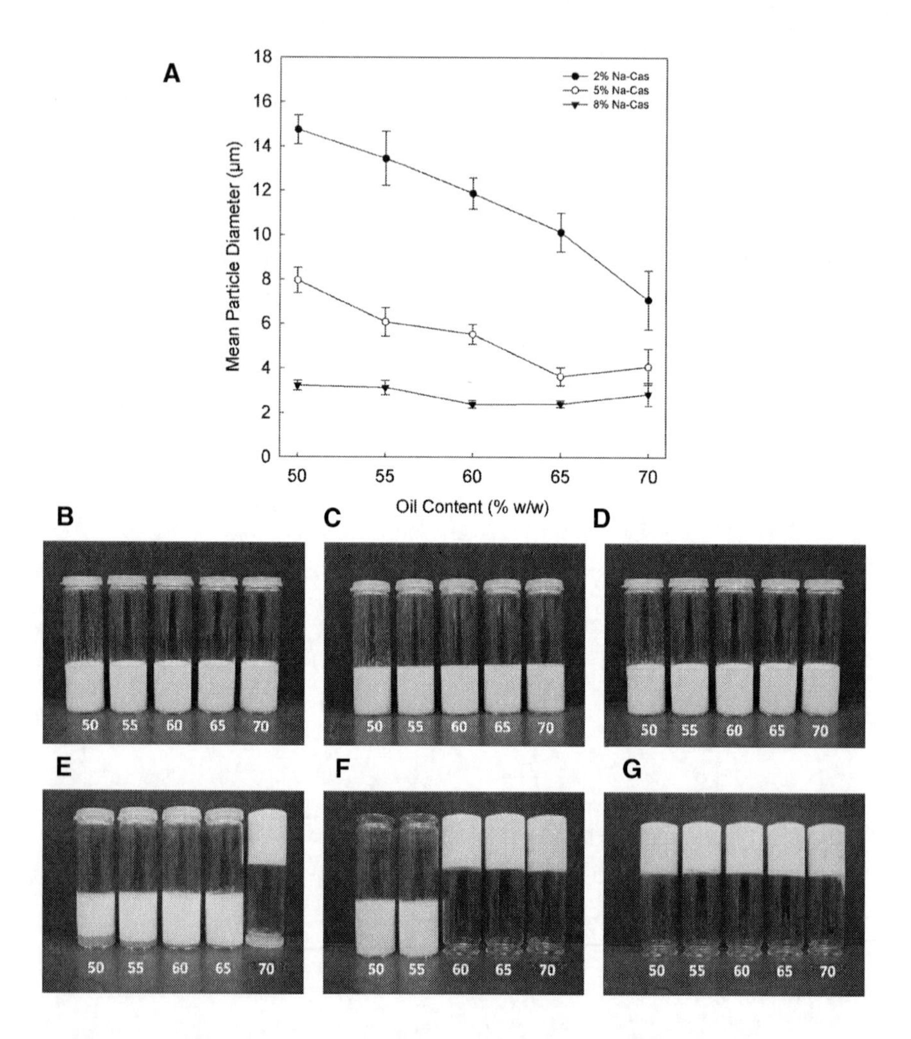


**Figure 4 Mean particle diameter ($d_{43}$) of oil-in-water emulsions (A) produced by high shear blending (20000 rpm, 3 min) as well as photographic images of emulsion gels (55 - 70 % (w/w) with various protein concentrations before and after transglutaminase incubation (37 °C, 15 h, pH 6.8): (2 (B, E), 5 (C, F), 8 (D, G) % (w/w) protein).**

Results show that the mean particle diameter affects the packing of emulsion droplets regardless of packing type. In case of very small sized droplets (e.g. $d$ = 30 nm), the droplets remain separated by distances of several nanometers, in particular at low oil droplet concentrations regardless of droplet packing. An increase in particle size leads to a shift in mean surface distances towards higher oil droplet concentrations. At oil volume fractions of $\Phi > 0.6$ the mean surface distance between particles becomes less than a few nanometers which might facilitate enzymatically-induced crosslinking of adsorbed proteins. Above this oil volume fraction, a particulate gel structures might form after transglutaminase was added since caseinates are known to protrude from the interface of an oil droplet to a depth of 1 to 10 nm (*Fang et al.*, 1993).

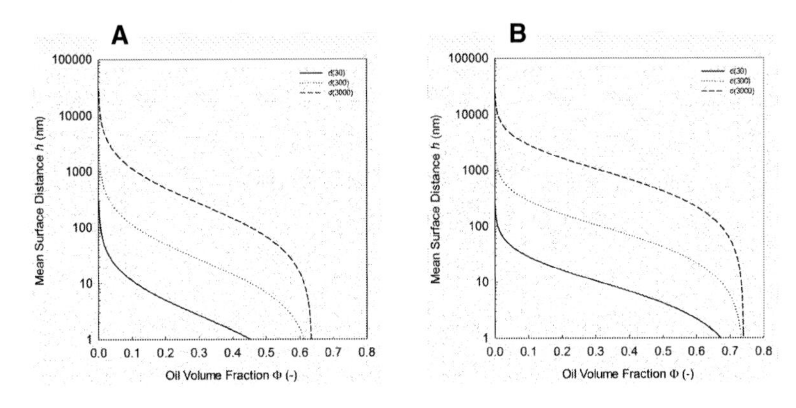


**Figure 5 Mean surface distance between droplets in emulsions (mean droplet diameter: 30, 300, 3000 nm) having a cubic (A) or hexagonal (B) packing.**

Moreover, **Table 4** shows results of a calculation of the time it takes a droplet of 200 nm to cross the mean separation distance – given by the oil content in a specific packing arrangement - by diffusion. There, the diffusion coefficient of droplets was calculated using the Stokes-Einstein equation (*Walstra et al.*, 2003):

$$D = \frac{kT}{6\pi r\eta} \qquad (5)$$

where $k$ is the Boltzmann Constant (1.38065 * $10^{-23}$ J/K), $T$ is the temperature of the system (295.15 K), $r$ the radius of the droplet (0.1 μm) and $\eta$ the viscosity of the surrounding phase (1 mPa*s). The average travel time $t$ of a droplet was then calculated assuming a diffusion-driven random walk (*Walstra et al.*, 2003; *Israelachvili*, 2010):

$$t = \frac{\overline{h^2}}{2D} \qquad (6)$$

Depending on the packing configuration, the intervening space between droplets can be crossed by diffusion in as little as 1 μs (**Table 4**). This means that droplet membranes are almost constantly in contact with each other giving transglutaminase sufficient time to initiate the formation of the covalent bonds between the functional groups.

**Table 4 Calculated average time between droplet-droplet collisions for emulsions having different oil droplet contents with droplet being arranged in a hexagonal or a cubic packing. Calculations were carried out assuming that droplets had a mean diameter of 300 nm and thus an effective diffusion coefficient of 2.162 × $10^{-12}$ $m^2/s$.**

| | hexagonal packing | | cubic packing | |
|---|---|---|---|---|
| oil (%) | surface distance (nm) | collision time (ms) | surface distance (nm) | collision time (ms) |
| 5 | 291.1 | 29.41 | 143.8 | 7.18 |
| 10 | 189.8 | 12.50 | 75.3 | 1.97 |
| 15 | 140.5 | 6.85 | 48.2 | 0.81 |
| 20 | 109.4 | 4.15 | 33.6 | 0.39 |
| 25 | 87.2 | 2.64 | 24.3 | 0.21 |
| 30 | 70.3 | 1.71 | 17.9 | 0.11 |
| 35 | 56.8 | 1.12 | 13.2 | 0.06 |
| 40 | 45.6 | 0.72 | 9.6 | 0.03 |
| 45 | 36.1 | 0.45 | 6.8 | 0.02 |
| 50 | 28.0 | 0.27 | 4.5 | 0.01 |
| 55 | 20.8 | 0.15 | 2.6 | 0.00 |
| 60 | 14.5 | 0.07 | 1.0 | 0.00 |

In addition, results have also shown that the overall concentration of protein present in the emulsion affects the formation of protein networks. We suggest that this is because of the ratio between adsorbed (bound) and excess (free) protein. Generally, when the interface of oil droplets is fully saturated, no more interfacially-active species can be adsorbed unless multilayers are formed. Thus, increases in surfactant concentration lead to them having an excess concentration in the aqueous phase (*Sanchez et al.*, 2005). Caseinate is known to have a surface protein concentration of about 0.5 - 1.36 $mg/m^2$ (*Hunt et al.*, 1994; *Srinivasan et al.*, 1996). In order to determine the non-adsorbed caseinate concentration in the surrounding water phase, we used the following equation:

$$c_a = \frac{\Gamma_S \, 6 \, \Phi}{d_{32}} \qquad (7)$$

where $c_a$ ($kg/m^3$) is the mass of emulsifier adsorbed to the surface of the droplets per unit volume emulsion, $\Gamma_S$ ($kg/m^2$) the surface load – assumed to be an average of 1 $mg/m^2$ (see above) - and $d_{32}$ (m) the volume-surface mean droplet diameter (*McClements*, 2004). Using equation (7), we estimated the adsorbed caseinate concentration of emulsions containing oil contents of 5 - 60% (w/w) and Na-Cas contents of 2, 5, and 8% (w/w), respectively (**Figure 6**). The aqueous caseinate concentration was calculated as:

$$c_w = c_i - c_a \tag{8}$$

where $c_w$ is the mass of Na-Cas (per volume of emulsions) in the aqueous phase and $c_i$ the mass of Na-Cas initially used.

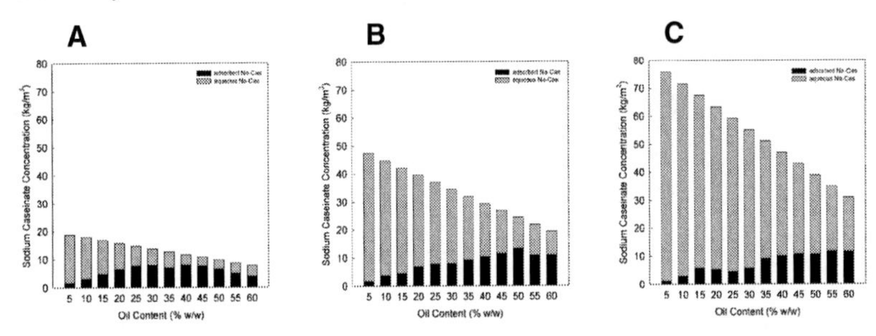


**Figure 6 Adsorbed (bound) and excess (free) aqueous Na-Cas concentration in oil-in-water emulsions having different oil contents (5 - 60% (w/w)) stabilized by 2 (A), 5 (B), and 8 (C) % (w/w) Na-Cas.**

The mass of emulsifier required to cover the droplet surface tended to increase with increasing oil volume fraction (*Hunt et al.*, 1994; *Srinivasan et al.*, 1996; *McClements*, 2004). Moreover, the concentration of non-adsorbed caseinate in the water phase remains quite high when the initial protein is high – a fact that might facilitate the transglutaminase-induced formation of a protein matrix with embedded oil droplets rather than the formation of a droplet-droplet network (**Figure 6**). Thus, the actual structural state of a particular protein-based emulsion gel might lie between the two extremes described above (*Dickinson*, 2012).

### Key insights

In summary, we can highlight a number of key insights obtained in our study, as illustrated in **Figure 7**:

- Transglutaminase-induced crosslinking of Na-Cas dispersions depends on the available protein concentration: Above a critical protein concentration, a Na-Cas network is formed, whereas below this concentration the dispersion remains liquid.

- Addition of transglutaminase to Na-Cas-stabilized emulsions can lead to (i) the emulsion remaining fluid or (ii) the emulsion undergoing a sol – gel transition. In the latter case, the gel that is formed may assume different structural arrangement i.e. a droplet network versus a protein network with embedded droplets. These systems have vastly different rheological properties.

- The formation of a protein embedded gel versus a droplet network depends on the oil droplet concentration which in turn affects the amount of free protein. At high oil droplet concentrations and low or intermediate protein concentrations, droplet-droplet networks are formed. In contrast, at high protein concentrations, protein-networks containing droplets are formed.

- The formation of a droplet network occurs only at critically high oil droplet concentrations > 60% (w/w) after addition of transglutaminase. This is because mean separation distances and time between droplet-droplet contact becomes sufficiently small.

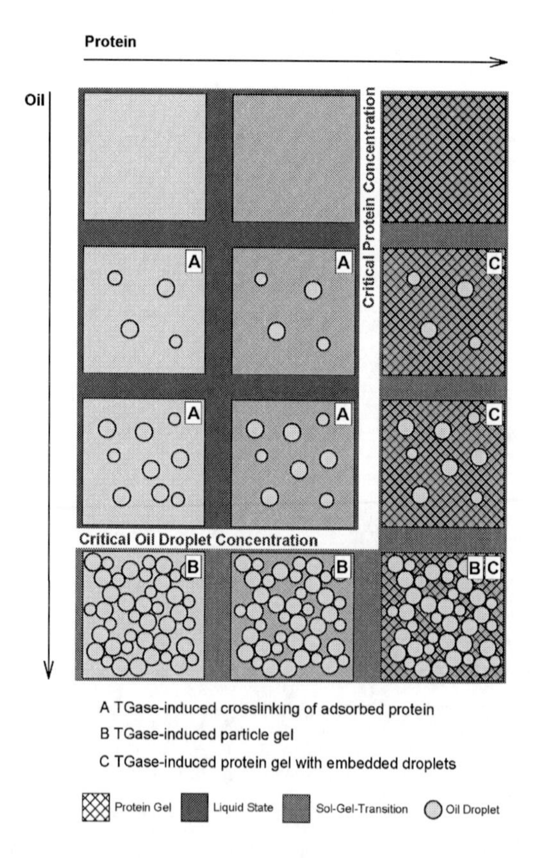


**Figure 7 Schematic mechanism of transglutaminase-induced gelation of oil-in-water emulsions depending on protein and droplet concentration.**

## CONCLUSIONS

This study demonstrated that enzyme-induced crosslinking is a useful tool to induce sol – gel transitions in protein-stabilized emulsions. The structure formed is influenced by both oil droplet and protein content. The influence of particle size on the formation of gel networks still remains to be investigated. Emulsions containing smaller droplets will generally have lower free protein contents and may thus be assumed to tend more to form droplet networks, especially at high oil contents. Mixing differently sized emulsions stabilized by different proteins and treating them with specific crosslinking enzymes may enable them to generate gel networks with specific functional properties. Furthermore, the physical state of the oil, temperature and enzyme activity may influence the structure created. As such, this study presents a first step in this direction. Further studies on this subject should be carried out to gain a better understanding on how to rational design such systems, particularly if they are to be used as delivery systems or texture modifiers.

## ACKNOWLEDGEMENTS

We would like to thank Rovita GmbH (Engelsberg, Germany) for generously providing us with sodium caseinate samples.

## REFERENCES

Adler-Nissen, J. (1979): Determination of the degree of hydrolysis of food protein hydrolysates by tri-nitro-benzenesulfonic-acid. Journal of Agricultural and Food Chemistry 27(6): 1256-1262.

Chanamai, R. and D. J. McClements (2000): Dependence of creaming and rheology of monodisperse oil-in-water emulsions on droplet size and concentration. Colloids and Surfaces A: Physicochemical and Engineering Aspects 172(1-3): 79-86.

Chen, J. and E. Dickinson (1998): Viscoelastic properties of protein-stabilized emulsions: Effect of protein-surfactant interactions. Journal of Agricultural and Food Chemistry 46(1): 91-97.

Chen, L., G. E. Remondetto, et al. (2006): Food protein-based materials as nutraceutical delivery systems. Trends in Food Science & Technology 17(5): 272-283.

De Jong, G. A. H. and S. J. Koppelman (2002): Transglutaminase catalyzed reactions: Impact on food applications. Journal of Food Science 67(8): 2798-2806.

Dickinson, E. (1997): Enzymatic crosslinking as a tool for food colloid rheology control and interfacial stabilization. Trends in Food Science and Technology 8(10): 334-339.

Dickinson, E. (2001): Milk protein interfacial layers and the relationship to emulsion stability and rheology. Colloids and Surfaces B: Biointerfaces 20(3): 197-210.

Dickinson, E. (2012): Emulsion gels: The structuring of soft solids with protein-stabilized oil droplets. Food Hydrocolloids 28(1): 224-241.

Dickinson, E. and Y. Yamamoto (1996): Rheology of milk protein gels and protein-stabilized emulsion gels cross-linked with transglutaminase. Journal of Agricultural and Food Chemistry 44(6): 1371-1377.

Færgemand, M., B. S. Murray, et al. (1997): Cross-Linking of milk proteins with transglutaminase at the oil-water interface. Journal of Agricultural and Food Chemistry 45(7): 2514-2519.

Færgemand, M., B. S. Murray, et al. (1999): Cross-linking of adsorbed casein films with transglutaminase. International Dairy Journal 9(3-6): 343-346.

Fang, Y. and D. G. Dalgleish (1993): Dimensions of the adsorbed layers in oil-in-water emulsions stabilized by caseins. Journal of Colloid and Interface Science 156(2): 329-334.

Folk, J. E. and P. W. Cole (1966): Transglutaminase: Mechanistic features of the active site as determined by kinetic and inhibitor studies. Biochim Biophys Acta 122((2)): 244-264.

Hinz, K., T. Huppertz, et al. (2007): Influence of enzymatic cross-linking on milk fat globules and emulsifying properties of milk proteins. International Dairy Journal 17(4): 289-293.

Hunt, J. A. and D. G. Dalgleish (1994): Adsorption behaviour of whey protein isolate and caseinate in soya oil-in-water emulsions. Food Hydrocolloids 8(2): 175-187.

Israelachvili, J. (2010): Intermolecular And Surface Forces, Academic Press.

Kamiya, H. and M. Iijima (2010): Surface modification and characterization for dispersion stability of inorganic nanometer-scaled particles in liquid media. Science and Technology of Advanced Materials 11(4).

Kellerby, S. S., S. G. Yeun, et al. (2006): Lipid oxidation in a menhaden oil-in-water emulsion stabilized by sodium caseinate cross-linked with transglutaminase. Journal of Agricultural and Food Chemistry 54(26): 10222-10227.

Kuraishi, C., K. Yamazaki, et al. (2001): Transglutaminase: Its utilization in the food industry. Food Reviews International 17(2): 221-246.

Lee, H. A., S. J. Choi, et al. (2006): Characteristics of sodium caseinate- and soy protein isolate-stabilized emulsion-gels formed by microbial transglutaminase. Journal of Food Science 71(6): C352-C357.

Littoz, F. and D. J. McClements (2008): Bio-mimetic approach to improving emulsion stability: Cross-linking adsorbed beet pectin layers using laccase. Food Hydrocolloids 22(7): 1203-1211.

Macierzanka, A., F. Bordron, et al. (2011): Transglutaminase cross-linking kinetics of sodium caseinate is changed after emulsification. Food Hydrocolloids 25(5): 843-850.

McClements, D. J. (2004): Food emulsions: Principles, practice, and techniques. Boca Raton, CRC Press.

McClements, D. J. (2004): Protein-stabilized emulsions. Current Opinion in Colloid & Interface Science 9(5): 305-313.

McClements, D. J., E. A. Decker, et al. (2009): Structural design principles for delivery of bioactive components in nutraceuticals and functional foods. Critical Reviews in Food Science and Nutrition 49(6): 577 - 606.

Mezzenga, R. and S. Ulrich (2010): Spray-dried oil powder with ultrahigh oil content. Langmuir 26(22): 16658-16661.

Minussi, R. C., G. M. Pastore, et al. (2002): Potential applications of laccase in the food industry. Trends in Food Science & Technology 13(6-7): 205-216.

Motoki, M. and K. Seguro (1998): Transglutaminase and its use for food processing. Trends in Food Science & Technology 9(5): 204-210.

Nonaka, M., H. Sakamoto, et al. (1992): Sodium caseinate and skim milk gels formed by incubation with microbial transglutaminase. Journal of Food Science 57(5): 1214-1241.

Partanen, R., A. Paananen, et al. (2009): Effect of transglutaminase-induced cross-linking of sodium caseinate on the properties of equilibrated interfaces and foams. Colloids and Surfaces A: Physicochemical and Engineering Aspects 344(1-3): 79-85.

Sanchez, C. C. and J. M. R. Patino (2005): Interfacial, foaming and emulsifying characteristics of sodium caseinate as influenced by protein concentration in solution. Food Hydrocolloids 19(3): 407-416.

Sharma, R., M. Zakora, et al. (2002): Characteristics of oil-water emulsions stabilised by an industrial Î±-lactalbumin concentrate, cross-linked before and after emulsification, by a microbial transglutaminase. Food Chemistry 79(4): 493-500.

Sharma, R., M. Zakora, et al. (2002): Characteristics of oil-water emulsions stabilized by an industrial □-lactalbumin concentrate, cross-linked before and after emulsification, by a microbial transglutaminase. Food Chemistry 79(4): 493-500.

Srinivasan, M., H. Singh, et al. (1996): Sodium caseinate-stabilized emulsions: Factors affecting coverage and composition of surface proteins. Journal of Agricultural and Food Chemistry 44(12): 3807-3811.

Tang, C. H., L. Chen, et al. (2011): Mechanical and water-holding properties and microstructures of soy protein isolate emulsion gels induced by $CaCl_2$, glucono-□-lactone (GDL), and transglutaminase: Influence of thermal treatments before and/or after emulsification. Journal of Agricultural and Food Chemistry 59(8): 4071-4077.

Velikov, K. P. and E. Pelan (2008): Colloidal delivery systems for micronutrients and nutraceuticals. Soft Matter 4(10): 1964-1980.

Walstra, P. and W. Walstra (2003): Physical Chemistry of Foods, Marcel Dekker Inc.

Woodcock, L. V. (1987): Developments in the non-newtonian rheology of glass forming systems. Molecular Dynamics and Relaxation Phenomena in Glasses. T. Dorfmüller and G. Williams, Springer Berlin Heidelberg. 277: 113-124.

Yang, M., F. Liu, et al. (2011): Properties and microstructure of transglutaminase-set soy protein-stabilized emulsion gels. Food Research International.

Zeeb, B., L. Fischer, et al. (2011): Cross-linking of interfacial layers affects the salt and temperature stability of multilayered emulsions consisting of fish gelatin and sugar beet pectin. Journal of Agricultural and Food Chemistry 59(19): 10546-10555.

Zeeb, B., M. Gibis, et al. (2012): Crosslinking of interfacial layers in multilayered oil-in-water emulsions using laccase: Characterization and pH-stability. Food Hydrocolloids 27(1): 126-136.

Zeeb, B., M. Gibis, et al. (2012): Influence of interfacial properties on Ostwald ripening in crosslinked multilayered oil-in-water emulsions. Journal of Colloid and Interface Science 367(1): 65-73.

# Concluding remarks and outlook

### Conclusion

This study has shown that crosslinking enzymes such as laccase or transglutaminase may be useful tools to covalently crosslink deposited biopolymers or single particles to improve the functionality of emulsion-based systems used as carriers for functional ingredients in terms of pH-, salt-, temperature-stability as well as Ostwald ripening and release kinetics. The composition, density, and thickness of the interfacial membrane has a major impact on the functionality of the derived biopolymer-based delivery systems. While the approach is working very well with e.g. multilayered emulsions where substrates are physically available to the crosslinking enzymes, crosslinking of biopolymers in aggregated structures such as soluble complexes, coacervates, and hydrogel particles is limited or even impossible since the enzyme is unable to access the substrate. In such structures, crosslinking may help to stabilize a base structure that can then later be assembled into another derived structure, but not the derived structure. This is illustrated in **Figure 1** that shows the relationship between *structure formation – structure modification – structure property*. In multilayered emulsions, enzymes are not sterically hindered and can thus access any biopolymer located at the surface of the emulsions. Likely, the approach can be translated to any other system where the target substrate is present at an accessible surface such as for example polymer-micelle complexes or biopolymer coated liposomes. Moreover, if emulsion droplet concentrations are high enough, droplet-droplet crosslinking may be induced leading to the formation of particulate networks with solid-like properties. Thus, this enzymatic approach can be used to modify stability, appearance, and organoleptic properties of food products – possibly even behavior during digestion (albeit additional studies would be needed to investigate this). The results thus should be of substantial interest to food manufacturer that are in need of having emulsions with improved functional properties. Taken together, the studies presented in this thesis show that the newly established science of biomimetrics – a new field that is under intense investigations in engineering, electronics and pharmaceutics – may be applicable to food science and technology. Use of enzymes in foods is in general much more advantageous than the use of harsh reactive chemicals, since they pose less problems in terms of allergencity or toxicity.

## Outlook

The quality of emulsion-based products is mainly influenced by their stability, rheology, and appearance (*McClements*, 2004). A physically induced colour change of oil-in-water emulsions was previously demonstrated by *Weiss et al.* (2001). As a consequence, enzyme technology might be useful to modify the appearance of emulsions as previously described in Chapter 5. Therefore, a more sophisticated experimental design has to be developed to further investigate these phenomena. Moreover, to date, relatively little is known about what happens when two enzymes are to be used simultaneously or in sequence on a target structures. It would be interesting to see whether deposited layers cannot only be intra-crosslinked, but also inter-crosslinked. For example, a crosslinking of membranes in a tightly packed system could lead to inter- rather than intra-crosslinking thereby creating particulate gel networks. Investigations need to confirm that two or more enzymes may be

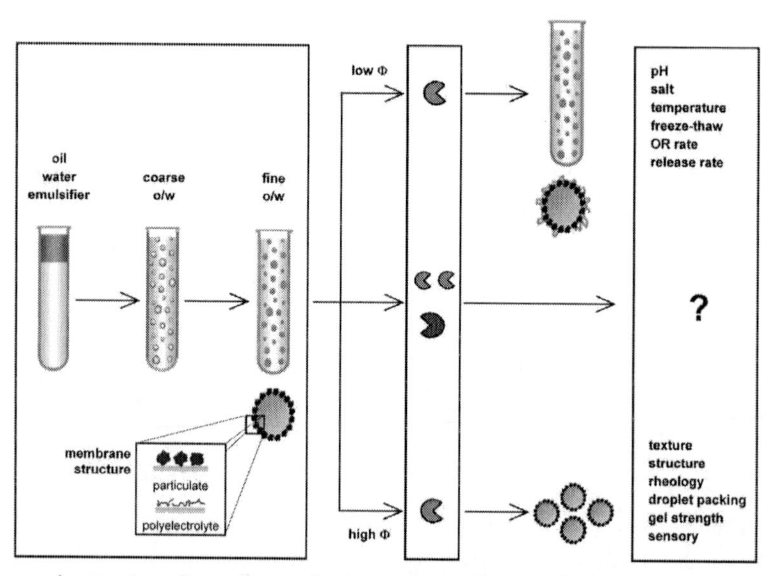


**Figure 1 Structure formation, modification, and properties after enzymatic treatments (crosslinking symbol = crosslinking enzyme; degrading symbol = degrading enzyme).**

used as suitable crosslinking agents in terms of pH-, temperature-, and salt-stability as well as kinetic parameters. Furthermore, the sequence of enzyme addition may play a major role in crosslinking proteins and/or polysaccharides in already assembled structures. Future studies may also focus on the use of degrading enzymes to specifically modify protein and/polysaccharide-based delivery systems. Earlier studies have already shown that trypsin was able to modify milk proteins before and after emulsification. It was demonstrated that emulsions formed from hydrolyzed protein solutions were less stable during storage than those formed from unmodified proteins and subsequently hydrolyzed or those formed with unmodified protein solutions (*Agboola et al.*, 1996). Moreover, the degree of trypsin-induced hydrolysis had a significant impact on the stability of oil-in-water emulsions exposed to different destabilization conditions (*Agboola et al.*, 1996). Another commonly used enzyme catalyzing the breakdown of chemical bonds is chymosin. As part of the renneting process, it destabilizes a milk protein-based colloidal system leading to an enzyme-induced gelation (*Dickinson*, 2012). Other proteinases have been successfully used to induce the formation of WPI gels (*Doucet et al.*, 2001). One distinct application of using degrading enzymes is the degradation of polysaccharides resulting in different molecular weight distributions which could be adsorbed to an interfacial membrane in order to modify the pore size of the polymer network. This may have an impact on the diffusion kinetics of encapsulated bioactives through the membrane. More recently, research focuses on the formation of $\beta$-lactoglobulin ($\beta$-lg) fibrils arised when $\beta$-lg is heated at pH 2 which might be used as thickening or gelling agents in food products (*Akkermans et al.*, 2008; *Kroes-Nijboer et al.*, 2011). Typically, aspartic acid residues in $\beta$-lg are more prone to an acid-induced cleavage at high temperatures; thus, enzymatic action might be used to cleave the same peptide bonds in the $\beta$-lg to promote the formation of fibrils (*Akkermans et al.*, 2008). Finally, future studies may also focus on what happens to enzymatically crosslinked structures during digestion. Possibly, such structures are more stable during digestion, which could offer benefits in terms of using them to deliver sensitive bioactives or probiotics or to induce a higher satiety feeling.

## REFERENCES

Agboola, S. O., Dalgleish, D. G. (1996): Enzymatic hydrolysis of milk proteins used for emulsion formation. 1. Kinetics of protein breakdown and storage stability of the emulsions. Journal of Agricultural and Food Chemistry 44(11): 3631-3636.

Agboola, S. O., Dalgleish, D. G. (1996): Enzymatic hydrolysis of milk proteins used for emulsion formation. 2. Effects of calcium, pH, and ethanol on the stability of the emulsions. Journal of Agricultural and Food Chemistry 44(11): 3637-3642.

Akkermans, C., van der Goot, A. J., et al. (2008): Formation of fibrillar whey protein aggregates: Influence of heat and shear treatment, and resulting rheology. Food Hydrocolloids 22(7): 1315-1325.

Akkermans, C., Venema, P., et al. (2008): Enzyme-Induced formation of β-lactoglobulin fibrils by AspN endoproteinase. Food Biophysics 3(4): 390-394.

Dickinson, E. (2012): Emulsion gels: The structuring of soft solids with protein-stabilized oil droplets. Food Hydrocolloids 28(1): 224-241.

Doucet, D., Gauthier, S. F., et al. (2001): Rheological characterization of a gel formed during extensive enzymatic hydrolysis. Journal of Food Science 66(5): 711-715.

Kroes-Nijboer, A., Venema, P., et al. (2011): Influence of protein hydrolysis on the growth kinetics of β-lg fibrils. Langmuir 27(10): 5753-5761.

McClements, D. J. (2004): Food emulsions: Principles, practice, and techniques. Boca Raton, CRC Press.

Weiss, J., McClements, D. J. (2001): Color changes in hydrocarbon oil-in-water emulsions caused by Ostwald ripening. Journal of Agricultural and Food Chemistry 49(9): 4372-4377.

# Summary

A biomimetic approach based on enzyme technology was used to modify physically-assembled food structures. Two different enzymes, namely an oxidoreductase (laccase) and an acyltransferase (transglutaminase), were used as crosslinking agents. First, the enzymatic crosslinking of biopolymer layers adsorbed at the interface of oil-in-water emulsions was investigated. A sequential two step process, based on the electrostatic deposition of pectin onto a fish gelatin interfacial membrane was used to prepare emulsions containing oil droplets stabilized by fish gelatin-beet pectin-membranes (*structure formation*). Laccase was added to the fish gelatin-beet pectin emulsions and emulsions were incubated for 15 min at room temperature (*structure modification*). The pH- and storage stability of primary, secondary (coated) and secondary, laccase-treated emulsions was determined. Results indicated that crosslinking occurred exclusively in the layers and not between droplets, since no aggregates were formed. Droplet size increased from 350 to 400 nm regardless of oil droplet concentrations within a matter of minutes after addition of laccase suggesting formation of covalent bonds between pectin adsorbed at interfaces and pectin in the aqueous phase in the vicinity of droplets. During storage, size of enzymatically treated emulsions decreased, which was found to be due to enzymatic hydrolysis (CHAPTER 2). Crosslinked pectin-coated oil droplets had similar or significantly better stability ($p < 0.05$) than oil droplets of primary or secondary emulsions to NaCl addition (0 – 500 mM), $CaCl_2$ addition (0 – 250 mM), and thermal processing (30 – 90 °C for 30 min). Freeze-thaw stability and creaming behavior of enzyme-treated, secondary emulsions after two cycles (-8 °C for 22 h; +25 °C for 2 h) was significantly improved ($p < 0.05$) (CHAPTER 3). Furthermore, the influence of interfacial crosslinking, layer thickness, and layer density on the kinetics of Ostwald ripening in multilayered emulsions at different temperatures was investigated. Growth rates of droplets were measured by monitoring changes in the droplet size distributions of 0.5% (w/w) *n*-octane, *n*-decane, and *n*-dodecane oil-in-water emulsions using static light scattering. Lifshitz-Slyozov-Wagner theory was used to calculate Ostwald ripening rates. Ripening rates of single-layered, double-layered, and crosslinked emulsions increased as the chain length of the *n*-alkanes decreased. Emulsions containing crosslinked fish gelatin-beet pectin coated droplets had lower droplet growth rates ($3.1 \pm 0.3 \times 10^{-26}$ $m^3/s$) than fish gelatin-stabilized droplets ($7.3 \pm 0.2 \times 10^{-26}$ $m^3/s$), which was attributed to the formation of a protective network (CHAPTER 4). Multilayering and enzymatic crosslinking of emulsions caused alterations in the release behavior of an encapsulated core material due to changes in thickness, porosity and permeability of the membrane. Release behavior of primary, secondary (coated), and laccase-treated secondary emulsions carrying lutein, an oxygenated carotenoid, were also characterized and studied. Primary WPI–stabilized emulsions showed a five times higher release of lutein after 48 h than secondary emulsions (pH 3.5). Primary DTAB-stabilized emulsions released 7.2% of encapsulated lutein within the observation period, whereas beet pectin-DTAB coated emulsions released only 0.13% of lutein. Crosslinking of adsorbed pectin did not significantly decrease release of lutein in comparison to non-crosslinked secondary emulsions (CHAPTER 5) (*structure properties*). Oil droplets of stable emulsions with different interfacial membrane compositions were also subjected to enzymatic crosslinking. The pH-stability of primary emulsions and nanoparticle-coated base emulsions was determined before and after laccase addition. However, results indicated that crosslinking occurred only in deposited monolayers rather than between adsorbed biopolymer nanoparticles (CHAPTER 6).

Moreover, the influence of oil volume fraction of protein-stabilized oil-in-water emulsions on the formation of cold-set particulate gels using microbial transglutaminase was assessed. Texture profile analysis and rheological measurements indicated that a critical oil volume fraction ($\Phi > 0.6$) was crucial to promote a transglutaminase-induced crosslinking of single oil droplets (CHAPTER 7).

# Zusammenfassung

Der Einsatz vernetzungsfähiger Enzyme wurde am Beispiel einer Laccase bzw. Transglutaminase zur Stabilisierung und Modifizierung neuer Verkapselungssysteme untersucht. Ziel der Studien war es insbesondere den Wirkmechanismus der vernetzenden Enzyme in komplexen Lebensmittelmatrizen zu klären. Im Fokus der Forschung stand die Wirkung der Oxidase Laccase auf eine Fischgelatine-Zuckerrübenpektin-stabilisierte Öl-in-Wasser-Emulsion (d = 350 ± 50 nm). In einem zweistufigen Prozess basierend auf der elektrostatischen Anziehung entgegengesetzt geladener Biopolymere wurden zunächst multilamellare Grenzflächenmembranen erzeugt und anschließend einer Laccase-induzierten Vernetzung unterzogen. Die Ergebnisse belegen, dass ausschließlich Biopolymere, die an der unmittelbar zur kontinuierlichen Phase hin zugänglichen Grenzfläche der Öl-in-Wasser Emulsionen durch die Laccase vernetzt werden konnten. Bildmikroskopische Analysen zeigten, dass in verdünnten Systemen eine enzymatisch-induzierte Aggregation einzelner Ölpartikel nicht stattfand. ζ-Potential- sowie Messungen der Partikelgrößenverteilung mittels dynamischer Lichtstreuung ergaben Änderungen des mittleren Partikeldurchmessers von 380 auf 290 nm über eine Lagerdauer von 10 Tagen. Die Ursache dieses Phänomens konnte auf eine Nebenaktivität des Enzympräparats selbst zurückgeführt werden, die zu einer Hydrolyse der adsorbierten Pektinschicht führte (KAPITEL 2). In pH-Titrationsexperimenten (pH 3,5 - 10) konnte gezeigt werden, dass eine Vernetzung der aus Pektin bestehenden Grenzflächenmembranen den Umgebungsbedingungen standhielt und damit die pH-Toleranz der Öl-in-Wasser Emulsion erheblich verbessert wurde. Kovalent-vernetzte Grenzflächenmembranen wiesen darüber hinaus gleiche bzw. signifikant höhere Stabilitäten ($p < 0.05$) gegenüber NaCl- (0 – 500 mM) und $CaCl_2$- (0 - 250 mM) Zugabe auf. Verringerte Koaleszenz-Erscheinungen ließen sich in Hitze- (30 – 90 °C für 30 min) bzw. Gefrier-Tau-Experimenten (-8 °C für 22 h; +25 °C für 2 h) beobachten (KAPITEL 3). Die Ausbildung dicker, widerstandsfähiger Grenzflächenmembranen um Öltropfen erhöhte zudem den Stofftransportwiderstand und konnte damit ein Verlangsamen molekularer Diffusion von *n*-Alkanen ermöglichen. Hierzu wurden zeitliche Änderungen der Partikelgrößenverteilungen von *n*-Oktan-, *n*-Dekan- und *n*-Dodekan-in-Wasser Emulsionen mittels statischer Lichtstreuung verfolgt. Basierend auf der Lifshitz-Slyozov-Wagner-Theorie konnten Ostwaldreifungsraten in Abhängigkeit der Membrandicke bzw. –dichte bestimmt werden. Die Ergebnisse zeigten, dass Laccase mehrschichtige Emulsionen hinsichtlich temperaturabhängiger Ostwald-Reifungsvorgänge stabilisieren kann. Die Reifungsraten von beispielsweise Dodekan-in-Wasser Emulsionen verringerten sich von 7,3 ± 0,2 x $10^{-26}$ $m^3/s$ (einschichtig) auf 3,1 ± 0,3 x $10^{-26}$ $m^3/s$ (zweischichtig-vernetzt) (KAPITEL 4). Neben der Grenzflächendicke und -zusammensetzung beeinflusst die enzymatische Vernetzung adsorbierter Biopolymere auch die Freisetzung von verkapselten, bioaktiven Komponenten. Am Beispiel einer Lutein-haltigen Öl-in-Wasser Emulsion konnte gezeigte werden, dass einschichtige Emulsionen nach 48 Stunden zehnmal mehr Lutein freisetzen als zweischichtige bzw. vernetzte Emulsionen, wobei die Lutein-Freisetzung generell vom pH-Wert abhing (KAPITEL 5). Der Aufbau komplexer Grenzflächenmembranen sollte Aufschluss darüber geben, ob sich Emulsionen hinsichtlich ihrer pH-Stabilität nach Enzymzugabe optimieren lassen. Stabile Emulsionstropfen mit verschiedenen Membrankomplexen (Partikulär vs. Polyelektrolytisch) wurden dazu hergestellt und ebenfalls einer Enzymbehandlung unterzogen. Dabei stellte sich heraus, dass kein Vernetzen der adsorbierten Nanopartikel stattfand und die behandelten Emulsionen nicht belastungstoleranter als die Referenzemulsion war (KAPITEL 6).

In konzentrierten Emulsionssystemen können vernetzende Enzyme neben der Katalyse kovalenter Bindungen innerhalb der Grenzflächenmembran auch Tropfen-Tropfen-Netzwerke induzieren. Die Ausbildung partikulärer Emulsionsgele hängt dabei wesentlich vom mittleren Abstand der Emulsionstropfen ab. Es zeigte sich, dass eine bestimmte minimale Kollisionsfrequenz in Abhängigkeit der Tropfenkonzentration notwendig war, um enzyminduzierte kovalente Bindungen auszubilden. Die Acyltransferase Transglutaminase katalysierte dabei die Vernetzung der Öltropfen ($d_{32}$ = 300 nm) ab einer kritischen Öltropfenkonzentration von 60% (KAPITEL 7).

Diese Dissertation hat gezeigt, dass vernetzende Enzyme wie Laccase oder Transglutaminase in der Lage sind, die Struktur und Eigenschaften von Lebensmitteldispersion zu modifizieren. Die Funktionalität dieser emulsionsbasierten Systeme als Träger für potentielle funktionelle Komponenten kann in Bezug auf pH-, Salz- und Temperatur-Stabilität oder Ostwaldreifungs- bzw. Release-Raten gezielt optimiert werden. Da die räumliche Verteilung der Biopolymere im System einen wesentlichen Einfluss auf die Wirkung der Enzyme hat, spielt der Zeitpunkt der Enzymzugabe im Prozess und die damit verbundene Vernetzung eine wichtige Rolle. Beispielsweise ermöglichen offene, physisch zugängliche Strukturen die Ausbildung kovalenter Bindungen. Eine weitere strukturelle Modifikation, die zu komplexeren, unzugänglicheren Strukturen wie Aggregate oder Netzwerke führt, muss nach der enzymatischen Behandlung induziert werden: *Strukturaufbau – Strukturmodifikation – Struktureigenschaften*. Zusammenfassend kann gesagt werden, dass der Einsatz von vernetzenden Enzymen als umweltfreundliche und gleichzeitig wirtschaftlich machbare Option genutzt werden kann, um Lebensmittel mit verbesserten physikalischen, funktionellen, rheologischen oder sensorischen Eigenschaften herzustellen.